P9-ELD-780

The Alberta Oil Patch
Then & Now

MAGIC LIGHT PUBLISHING
Ottawa – Vancouver

The Alberta Oil Patch
Then & Now

By: John McQuarrie
Photography: John McQuarrie

Published by: Magic Light Publishing
                 John McQuarrie Photography
                 192 Bruyere Street
                 Ottawa, Ontario
                 K1N 5E1

                 (613) 241-1833
                 FAX: 241-2085
                 e-mail: mcq@magma.ca

Design: John McQuarrie
Printing: Paramount, Book Art, Hong Kong

Library and Archives Canada Cataloguing in Publication

McQuarrie, John, 1946-
        The Alberta oil patch : then & now / John McQuarrie.

ISBN 978-1-894673-43-3 (bound).--ISBN 978-1-894673-44-0 (pbk.)

        1. Petroleum industry and trade--Alberta--History. 2. Petroleum industry and trade--Alberta--History--Pictorial works. I. Title.

HD9574.C23A54 2010   333.8'232097123   C2010-904114-3

Front cover:
Back cover: Aerial view of Shell's Millennium Mine north of Fort McMurray, Drilling at the Vermillion oil field ca. 1938 (Page 8) and Modern, coil-tubing rig near High River (Page 9)

Title page: Motorman, Brandon Erdos and roughneck, Tyler Rook working the drilling floor of Nabors rig #19 near High River.

CONTENTS

This book was to have been a simple look at the history of the oil and gas industry in Alberta from its humble beginnings through to today. Things started quite inauspiciously in 1901 with the first oil discovery on Cameron Creek in today's Waterton Lakes National Park. The word "discovery" might be a bit of a misnomer here as the First Nations people in the area had known about it for centuries. While the oil here was too deep to recover with the primitive equipment of the time, the oil exploration seed had been planted and drilling rigs became a common sight throughout the province.

But it would be more than ten years later that everything changed when Alberta's first producing well, Dingman #1 came in. It was 1914 and overnight, the sleepy little town of Turner Valley south of Calgary, was transformed into a boom town and the economy of Alberta was forever changed. After the Second World War, as the Turner Valley reserves were beginning to dry up, Imperial Oil's Leduc #1 came in and a vast new reserve was discovered just south of Edmonton.

Through the last half of the twentieth century something called the oil sands slowly emerged as the new giant in Alberta's oil and gas industry and the story became even more interesting – and just a little bit complicated.

Pumpjack in the heart of Turner Valley.

OUR WORLD NEEDS ENERGY AND CANADA
IS UNIQUELY POSITIONED TO PROVIDE IT,
RESPONSIBLY AND RELIABLY.

Nabors, coil tube drill rig #88 bathed in the first light of an Alberta sunrise near Innisfail.

INTRODUCTION

Our world needs energy and Canada is uniquely positioned to provide it – responsibly and reliably. Oil and gas will continue to play a significant role in the global energy supply mix.

The Canadian oil and gas industry fully recognizes that it must continue to do its part in addressing greenhouse gas (GHG) emissions and it advocates several key principles to guide the development of Canadian climate policy.

The world needs energy. Lots of it. This need has positive implications for Canada's economy because our country is uniquely positioned to provide safe and secure energy – all kinds of energy – to customers across North America.

Energy production, from oil sands to renewables, must increase to meet growing demand. However, with energy demand in developing nations increasing, renewables alone won't keep pace. All sources of energy, including natural gas and crude oil, are needed. Canada is the only OECD (Organization for Economic Co-operation and Development) country with growing oil production, which means more jobs and investment. For example, oil and gas currently provides jobs for 500,000 Canadians, and that number is expected to grow.

By most measures, Canada consistently ranks among the top 10 energy producers in the world. Our production exceeds domestic consumption and we are the largest supplier of oil and gas to the U.S. Canada has more stringent environmental regulation than many other suppliers to the U.S. and our policies and regulations continue to evolve.

CANADA IS:
Second in the world in **crude oil** reserves.
Third in the world in **natural gas** production.
First in the world in **uranium** reserves.
Second in the world in **hydro-electricity** generation.

The value of Canada's energy exports in 2008 was $126 billion, accounting for 6 per cent of the country's GDP and 26 per cent of all Canadian exports.

Canadian Association Of Petroleum Producers   www.capp.ca

But it wasn't always so . . .

IN THE BEGINNING

Ironically, the recorded history of fossil fuels in Western Canada began in 1719 when Wa-pa-su, of the Cree First Nations, brought a sample of the oil sands to the Hudson Bay post at Fort Churchill. The sands were described as "that gum or pitch which flows out of the banks of that river." Fifty years later Peter Pond (1740-1807), the first European to enter the rich Athabasca fur-trading region (heart of today's oil sands), described the heavy oil outcroppings along the river, noting the Aboriginal People's use of the material to waterproof canoes. A Connecticut Yankee, Pond became a giant in the Canadian fur trade and is credited with inspiring Alexander Mackenzie to become the first white man to reach the Pacific Ocean overland across North America in 1793. This was 12 years before Lewis and Clark. The circumstances that brought these two men together are worth the telling.

Bill McDonald (www.peterpondsociety.com) describes their relationship. "A rough, intimidating, hulk of a man with a short temper, Pond was held in awe by both white men and Indians. He made the first crude maps of North America west of Hudson Bay and was a founding partner of the Northwest Company that faced vicious competition with the Hudson Bay Company before the latter swallowed the former in 1820. Pond was implicated in two murders in the wild north, one for which he was never convicted, the other committed by followers but for which he was blamed. That murder got him replaced in his post by Alexander Mackenzie to whom he had to explain the surrounding tribes and terrain. One aspect intriguing Mackenzie was Pond's belief that a nearby river stretched all the way to the Pacific, the likely mythical Northwest Passage."

Colourful characters make regular appearances in the pages of Alberta's rich oil history and another of these eccentric individuals, Kootenai Brown, is no exception. John George Brown (1839-1916) was an Irish-born Renaissance Man soldier, trader and conservation advocate. After serving as an army officer in India in 1858-1859, he sold his commission and joined the flood of prospectors flocking to the Cariboo Gold Rush. He proved unsuccessful as a prospector, turning to trapping, whisky trading, riding pony express, scouting for General Custer, bison hunting, wolfing and then briefly policing, serving as constable in Wild Horse Creek, BC. Quite a resumé! In 1874, he married a local Metis woman and, that same year, obtained a commission from the Geological Survey to find oil. With the help of the indigenous peoples, he was successful. By dumping gunny sacks into the seeps at Oil Creek, squeezing the oil into containers and selling bottles to local farmers for use as a lubricant, he assured himself a place in history by becoming the first person known to have marketed a petroleum product in Alberta.

MILESTONES

"In 1883 a CPR crew drilling for water was surprised to find natural gas. The rig caught fire and injured a number of the men. This was the first gas well in Alberta.

In 1888, one of Alberta's first cowboys, John Ware, was riding west of Okotoks when he stopped at a pond to water his horse. The animal refused to drink and he noticed a light film on the water emitting a sulphurous odour. When he tossed a lit match onto the surface, the resulting explosion frightened his horse into the brush. John Ware and Sam Howe are credited for discovering an oil seepage on the Stoos farm in Turner Valley.

A drilled well that gave Alberta its first oil strike was indicated by the surface shows of oil at Cameron Creek. The Rocky Mountain Development Company, incorporated by A.P. Patrick, J. Leeson and J. Lineman, found oil in 1902 at a depth of 312 metres. A flow of fine crude oil commenced with an estimated rate of 300 barrels/day. Rocky Mountain Development #1 was western Canada's first producing oil well. That initial strike was enough to launch Alberta's first oil boom, inspiring other oil companies to come to this field which became known as Oil City. Because of limited production, this did not become Alberta's first commercial field as the formation was an anomaly. Drilling activity greatly declined by 1905.

The first major gas field discovery in Alberta was made by Eugene Coste at Bow Island, southwest of Medicine Hat. Drilling of old Glory #1 began in 1908. A large natural gas field was found in 1909 which launched Western Canada's natural gas industry and a pipeline was constructed. It was Canada's first major pipeline, built with 16 inch pipe. It traveled southwest of Medicine Hat to small communities along the way, ending with Calgary in 1912, a distance of 274 kilometres. The pipeline was mainly hand dug with picks and shovels through rock, gravel and hard packed soil. Little in the way of mechanized equipment was used.

In the early days the rough and tumble working conditions contributed to serious injury and death. Roughnecks began with little experience, and received their training on the job, in the school of hard knocks. Safety training and operational procedures were limited. Twelve-hour shifts and seven day work weeks created exhaustion that contributed to dangerous errors. Wellhead pressure, poisonous fumes, moving equipment, and speed also led to deadly conditions.

The turning point in Canada's exploration effort came with the discovery in 1914 of the province's first major oil field. This milestone event in Turner Valley led to an oil and gas boom in which over 500 oil companies were formed in one day."

"Dingman #1 became the cornerstone of the province's young oil and gas industry.

During the 1920s, large corporations and small companies alike searched for petroleum, using a combination of intuition, good luck and surface geology. Geological clues had to be wrested from the clutches of muddy roads, frigid temperatures, primitive equipment, poor training and lack of money, supplies and support. Despite such difficulties, hope and optimism were in good supply and investors continued to support oil exploration.

By the end of the 1940s ninety percent of all crude oil refined in Canada had to be imported.

The modest wet gas find at Turner Valley in 1914, with fuel suitable to pour directly into the gas tanks of vehicles, was the precursor of the nearby deeper zone find of 10 years later.

The 1924 discovery of Royalite #4 launched the area south of Calgary soundly into the Canadian oil industry consciousness. The third oil strike was completed in June of 1936, with the Turner Valley Royalties wildcat well kicking off another boom, allowing Turner Valley to retain the status of the largest oil field in Canada.

Because of a lack of markets, many Turner Valley operators flared off millions of dollars worth of gas. This damaged the reservoir by reducing the pressure, resulting in diminished oil production from many wells. In 1938, the Petroleum and Natural Gas Conservation Board was created to try and ensure that Alberta's petroleum resources were well managed and used for public interests.

Throughout every decade of the twentieth century, the Athabasca Oil Sands continued to attract the attention of federal, provincial and independent explorers who were trying to find the answer to the sticky problem of how to profitably extract oil from the world's largest known petroleum deposit.

Even more money was spent on drilling some 1,000 exploratory wells for conventional oil in Alberta, which did not produce any large oil fields.

With the outbreak of WWII Canada needed as much petroleum as the industry could provide. Petroleum consumption, particularly of gasoline, had soared. Aviation fuel came from Turner Valley crude refined in Calgary. Vehicles building and travelling on the Alaska Highway needed fuel and they consumed thousands of barrels.

Canada could not meet the growing domestic demand for petroleum products, and by the end of the 1940s, ninety percent of all crude oil refined in Canada was imported. Turner Valley's declining reserves saw oil production spiral downward after 1942. Total production at the end of 1946 stood at seven million barrels while Canadian consumption was over 10 million.

Most petroleum workers had little formal training. Provincial certificates were required only for steam engineers. Most workers learned the necessary skills on the job. Fortunately, the end of WWII brought home men more skilled than many of the earlier workers.

On February 13, 1947 drillers at Imperial Leduc #1 discovered a huge oil field. This opened up the district and the region to tremendous oil exploration and petroleum discoveries.

Alberta's prosperity was built by ordinary people with exdtraordinary hope and strength. The province's identity, history and fame are intimately bound to her oil and gas industry – a vital part of her socio-economic life-blood. While the industry often makes news, there is seldom mention that it was generations of people who developed these resources and built the infrastructure we enjoy today.

Petroleum sector facts and figures may describe the development of the industry and its importance to Alberta's fortunes over the last 100 years but the history is not complete without including the hard-working people who made their contributions. It is the stories of the people who participated in the industry that lend life and colour to its history.

Perhaps unnoticed are the workers' families who moved with the breadwinner to the next job site, often enduring similar conditions as the workers. Today the resource industry is still a challenging environment for its workers and their families. Many people spend long shifts in the harsh environment of the oil camps and work sites of the supporting industries."

www.soulofalberta.com

SOUL OF ALBERTA IS AN ORGANIZATION DEDICATED TO
THE DESIGN, DEVELOPMENT AND DELIVERY OF
MEANINGFUL PRODUCTS AND SERVICES THAT
HELP CANADIANS TO DISCOVER OUR HISTORY,
VALUE OUR ACHIEVEMENTS
AND ENRICH OUR COMMUNITIES.

Oil workers operate canner #1 well at the Vermillion oil fields ca. 1938, Borradaile, Alberta.
(Canada Science and Technology Museum, CN003638) www.sciencetech.technomuses.ca

Oil workers on floor of Nabors Drilling rig #51, High River. Alberta. (Left to right, Greg Elliott, Ron Bailer and Jason Moore

WATERTON LAKES

Prince of Wales Hotel in its magnificent setting above Waterton Lake.
The large photo above is the view from the hotel's lawn.

Cairn in the shape of an oil rig marks the historic site of this pioneer development in Alberta's oil industry.

A handful of visitors to Waterton came to make their fortunes, not in gold or silver, but in the oil that the natives had known about for a long time. Eventually, an enterprising group decided to tap the untold wealth. The Rocky Mountain Development Company commenced drilling operations along Cameron Creek in November 1901, finally hitting oil at 311 metres by September of 1902, the first producing oil well in western Canada. Oil City was born – with rigs, a bunkhouse and dining hall, cabins and even the beginnings of a small hotel. Original Discovery #1 broke down and, when it was finally repaired, ceased to yield oil. Nor could the company find oil anywhere nearby. So Oil City was abandoned by 1907. A strike near Cameron Falls yielded something in the neighbourhood of one barrel a day until the well walls collapsed.

Waterton Lakes National Park likely owes its existence to the shortcomings of early oil exploration technology. The primitive drilling equipment could not penetrate the resistant rock of the Lewis Overthrust. The producing wells of today, located north of the park, hit oil below 6,000 feet. Had we the technology then that we have now, Waterton Lakes might be a vast oilfield instead of a National Park.

The site has since been reclaimed and all that remains is a cairn in the shape of an oil rig marking the historic site of this pioneer development in Alberta's oil industry. Oil City was designated as a national historic site in 1968.

www.greatcanadianparks.com

(Glenbow Archives, NA-4089-2)

Cameron Creek in flood, 1919. Waterton National Park. Original well is the one to the right.

Just a few minutes drive to the west of the Prince of Wales Hotel, this couple is exploring the area occupied by Oil City on Cameron Creek.

HIGHWAY 22

If you were compiling a list of the ten top drives in Canada, the 160-kilometre (100-mile) stretch of highway 22 known as the Cowboy Trail between Calgary and Pincher Creek (another 20 kilometres to the east on Highway 3), would be right up there near the top. The last 20 kilometres of this journey on Highway 3 offers is a number of dramatic vistas of wind farms revealed to you later on in your journey through this book. It also takes you right through Turner Valley, site of one of the most pivotal points in Alberta's oil history.

A pumpjack (also known as nodding donkey, oil derrick, pumping unit, horsehead pump, beam pump, sucker rod pump, grasshopper pump, thirsty bird and jack pump) is the overground drive for a reciprocating piston pump installed in an oil well.

It is used to mechanically lift liquid out of the well if there is not enough bottom hole pressure for the liquid to flow all the way to the surface. The arrangement is commonly used for onshore wells producing relatively little oil. Pumpjacks are common in many oil-rich areas, dotting the countryside and occasionally serving as local landmarks.

Depending on the size of the pump, it generally produces 5 to 40 litres of liquid at each stroke. Often this is an emulsion of crude oil and water. The size of the pump is also determined by the depth and weight of the oil to be removed, with deeper extraction requiring more power to move the heavier lengths of sucker rods.

A pumpjack converts the rotary mechanism of the motor to a vertical reciprocating motion to drive the pump shaft, and is exhibited in the characteristic nodding motion. The engineering term for this type of mechanism is a walking beam. It was often employed in stationary and marine steam engine designs in the 1700s and 1800s.

Pumpjack amidst sea of canola.

The Calgary Albertan, June 25, 1937

Today a visitor to the new South Field in Turner Valley is confronted with a bristling forest of sky-reaching derricks spreading between the hills. Steel derricks, wooden derricks, piles of pipeline, mounds of muck, pools of dark muddy water with a trace of oil. Dotting this man and machine-made panorama fiery flares from producing wells add to the unreality of the scene.

The motorist drives right into the main street of Little Chicago. The road is flanked with restaurants, lumber yards, stores of every description, a dance hall, movie theatre, pool rooms, lodging houses, bachelor shacks, trailers and even tents. More than one hundred buildings, not all painted, have sprung up like mushrooms almost overnight, and every day adds another one. They are not ornate, nor architecturally perfect, but all are sufficient to meet the needs of many hundred oil field workers.

The visitor is sure to ask "What are all those fires?" for in every backyard in Little Chicago, Little New York and even in Hartell the ground spurts flame from small sunken spots. A close look reveals tin can dumps, for these holes house natural gas incinerators. There are no garbage collectors needed here. A network of pipes that snake the surface of the fields with little outlets everywhere takes care of refuse and waste so that just the blackened tins remain.

The motoring tourist will have days of beautiful mountain views before reaching Calgary, but nowhere in the British Empire will his eyes behold a sight like "The Valley" with its flares, its derricks, its booming, relentless activity. The romantic story of discovery is unforgettably illustrated by that "Little Chicago" town. It's a trip all should plan to make and an experience none will live long enough to forget. You don't know Canada until you have witnessed the sight of Turner Valley's oilfield wonders.

(Glenbow Archives NA-67-83)   View of Little Chicago-Royalties, Turner Valley oil field, ca 1930 and this same area as it appears today (right).

TURNER VALLEY

South Turner Valley oilfield, June 28, 1937. (Provincial Archives of Alberta, P1800) It is interesting to note that the two photos on the previous pages were both taken from the hill seen in the background here.

"Longtime Turner Valley resident Evelyn Hayhurst has an unconditional love for the town. In 1939, her new husband Tom Hayhurst moved Evelyn into Turner Valley to start their life together. The couple stayed in Turner Valley, living on the corner of Sunset Blvd. and George Street for 70 years and running the Shop Rite grocery store on Main Street for 44 years.

When the town incorporated in 1930 and for many years following, it was a time when the night sky was lit up by the dozens of flares and the skyline was dominated by pumpjacks and wells.

Historian David Finch said the discovery of oil in Turner Valley in 1914 kickstarted the energy industry in Alberta and was the catalyst for a town to form. "Turner Valley is why all the big tall office towers with the names of oil companies on top are in Calgary and not Edmonton," said Finch, who wrote the book *Hell's Half Acre: Early Days in the Great Alberta Oil Patch*. "There were hundreds of oil wells drilled in the 1920s. Both Black Diamond and Turner Valley boomed. There were more people there then than there are now. There was even a daily airplane service between Calgary and Turner Valley, a commuter plane, because it was booming." The plane brought investors from Calgary out on a sightseeing tour of the oilfields, Finch said.

Dingman Discovery Well, 1914. (Provincial Archives of Alberta. P1301)

Precision Well Servicing rig #104 pulling pipe out of an old gas well in the Little Chicago area featured on the previous page. This is part of the process of an 'abandonment' that will eventually remove all evidence that there was ever a well here.

The Hayhursts' lives were connected to the light of the flares, the smell of the gas and the rhythm of the pumpjacks drawing oil out of the ground, particularly the pumpjack 50 feet from their house. "We could feel the well at night in our home, you could feel it pump," said Evelyn. "We got used to that. When they did close it in, cemented it in, it was funny. We missed it." The flare, the pumpjacks, the smell, it made Turner Valley unique and it made it feel like home. "When you got within three miles of Turner Valley you could smell it and we'd say, 'We're home' because you could smell the gas," said Evelyn, now 92 years old. She was fond of the flares, especially the flames along the Sheep River at the Turner Valley Gas Plant. "Oh, heaven's sakes there must have been 25 or 30 small flames," said Evelyn. "It was beautiful. Just lovely."

The oil and gas industry was the engine that drove the local economy and the town flourished.

"We used to dance once per week in the Legion hall," said Evelyn, now a widow and living in the High Country Lodge in Black Diamond. "It was fun. Turner Valley was a really going place."

Adapted from: *Turner Valley celebrates its historic past*, Okotoks Western Wheel
by: Tamara Neely

Precision Well Servicing rig #29 pulling pipe on a gas well near Black Diamond that has been producing since the 1920s.

Lowering bit at British Wainwright well, Turner Valley, Alberta, ca. 1928. (Canada Science and Technology Museum, CN000100)

It all started in Turner Valley – the birthplace of Alberta's oil and gas industry. In 1914 the first producing well came into production. It was known as Dingman #1.IA local rancher who had worked in the Pennsylvania oil fields, William Herron, noticed gas bubbling along the banks of Sheep Creek in the Turner Valley. Herron had some samples analyzed and found they contained hydrocarbons. He acquired more land in the area and later teamed up with Archibald Dingman from the Calgary Petroleum Products Company. They struck wet gas on May 14, 1914 while drilling the Dingman #1 well. This would be the first of three oil booms for the area. This one lasted only until August, 1914 when World War I started.

Few people at the time realized that the Dingman Discovery Well would change Alberta's economic future well into the 21st century. For 30 years, the Turner Valley Oilfields were a major supplier of oil and gas and the largest producer in the British Empire! The Turner Valley Oilfields became the cornerstone of Alberta's early oil and gas industry and the training ground for the industry as we know it today.

A remarkable history of Turner Valley's early years was marked by three major "booms" that occurred in this once famous oilfield. So significant was the role Turner Valley played in the history of the oil and gas industry that the Federal Government declared the Turner Valley Oilfields Gas Plant a National Historic Site.

The second oil boom was initiated in the spring of 1924 when the Royalite #4 well blew in producing 21 million cubic feet of wet gas and over 660 bbls of white naphtha per day. The Royalite #4 could not be controlled for weeks. There were no pipelines for the gas-making the oil the valuable commodity, while the gas was allowed to blow free with flares.

The third boom for the Turner Valley Oilfields came on June 16, 1936, with the discovery of crude oil. Robert Brown, an electrical engineer from Quebec, believed crude oil lay deep below the gas wells at Turner Valley. Brown formed the Turner Valley Royalties Company and, in 1936, the Turner Valley Royalties #1 struck oil and started a new era at the field and a third oil boom for the area. By 1939, the field had 70 oil wells producing over 10 million barrels of oil to assist in the World War II war effort.

Oil from the early wells was a clear naphtha with very low sulphur content, making it possible to be used in automobiles straight from the well.

Before pipelines were laid connecting the gas supply to Calgary, there wasn't a large market for the gas that was produced from the wells. It was the oil that was valuable. Consequently, the gas that was not used at the drilling site for heating homes in the Turner Valley area was allowed to blow free and burn. During this time, the Turner Valley skies were lit both day and night from the glow of the magnificent flares that burned off the gas. Turner Valley Oilfields became the largest oilfield in Canada and for more than ten years produced 200 million cubic feet of gas daily, enough to have supplied the daily needs of New York City.

Even when pipelines were eventually laid to Calgary, there was still far too much gas than was needed. Excess gas from the Royalite Oil Co. Gas Plant was piped from the plant and discharged into a ravine where it was burned. The area around this great flare became known as Hell's Half Acre. As stories would say, during the Great Depression, people would come from all over Canada and the U.S. looking for work in the Turner valley Oilfields. On cold winter nights, these travellers would huddle around the banks of Hell's Half Acre to keep warm.

Town of Turner Valley – www.turnervalley.ca

Precision Well Servicing rig #29 pulling pipe on a gas well near Black Diamond that has been producing since the 1920s.

TURNER VALLEY GAS PLANT

The gas glant constructed in Turner Valley during the 1930s remained operational until 1985. When abandoned, the provincial government acquired the site and began working in cooperation with the federal government on the preservation of the gas plant as a National Historic Site.

From within the bounds of the Town of Turner Valley is an impressive industrial complex of tanks, pipelines, domed buildings and scrubbing chimneys. This is the Turner Valley Gas Plant, the only surviving example of its kind in Canada and a pioneering component in one of the most important oil and gas fields in Alberta.

Although the gas plant no longer functions, this remarkable collection of structures, dating back to 1933, provides a historical study of the early developmental period in Alberta's oil and gas industry. It houses a considerable amount of intact oil-and-gas-processing equipment, much of which was state of the art and some of which was truly innovative when it was installed. The Turner Valley Gas Plant boasts Canada's first high pressure absorption gas extraction plant, first sour gas scrubbing plant (1935, 1941) and first propane plant in Canada 1949-1952). Other technological achievements include being one of Canada's first two sulphur plants (1952). It also includes remnants of distribution networks which employed both above and below ground pipelines. Such survivals largely unchanged, provide an important physical reminder of the complex processes necessary to refine and deliver gas and oil early in this century.

Town of Turner Valley – www.turnervalley.ca

But the industry now, ever since production halted in the mid 1980s, is mostly gone. It's an area that still carries its economic DNA – ranching/agriculture and petroleum production. With its quaint shops, artisan studios and breathtaking scenery, Turner Valley is a place for tourists and historians. In the shadow of tanks, pipelines, domed buildings and scrubbing chimneys, people now golf and fish and go to Rotary Club and the Foothills Figure Skating Club.

Turner Valley Gas Plant, National Historic Site today.

Turner Valley Gasworks, ca. 1930. (Glenbow Archives PAA 71.162)

Alberta has received international attention for its successes in natural gas flaring and venting reductions. Since 1996, solution gas flaring in Alberta has been reduced by 76 percent. In 2007, the upstream oil and gas industry conserved nearly 96 percent of all solution gas produced in Alberta for use and sale, rather than flaring and venting it. By reducing the amount of natural gas that is wasted, Alberta helps ensure the most effective use of natural gas resources and continues the province's record as one of the global leaders in conservation and flaring reductions.

www.energy.alberta.ca

"Flaring is the burning of natural gas that cannot be processed or sold. Flaring disposes of the gas while releasing emissions into the atmosphere. Most flaring performed in Western Canada involves "sweet gas" which is natural gas containing little or no hydrogen sulphide (H_2S).

Flare in Turner Valley oilfield. (Glenbow Archives, NA-67-86)

Flaring is also used to dispose of sour gas containing H_2S and waste gas containing contaminants such as H_2S and Carbon Dioxide (CO_2). It is a very important safety measure at natural gas facilities as it safely disposes of gas during emergencies, power failures, equipment failures or other "upsets" in the processing. Regulators have established guidelines for flaring reduction. Flare reduction increases the amount of marketable product being recovered and sold while also reducing emissions to the atmosphere.

These operation types include:

Solution Gas: Flaring   Natural gas contained in crude oil is called "solution gas". Flaring is used to dispose of natural gas produced along with crude oil and bitumen. If at all possible, this gas is recovered and pipelined to a processing facility. When oil is underground, the pressure of the reservoir holds gas in the oil and the pressure is reduced when the oil comes to the surface. This occurs at facilities called "batteries" where production from one or more wells is produced and stored.

Gas Plant Flaring: Gas processing plants remove the water, hydrogen sulphide, carbon dioxide and natural gas liquids from the raw natural gas to produce the market-ready natural gas. Flares are used to dispose of the unmarketable gases. All gas plants have flares to burn off gas safely during emergencies or "upset" conditions that interrupt the normal day-to-day operations. Many of the small plants are licensed to flare H_2S rich gas after it has been removed.

Well Test Flaring: Well test flaring occurs during drilling and testing of all oil and gas wells. This is a standard practice used to determine the types of fluids the well can produce, the pressure and flow rates of fluids and other characteristics of the underground reservoir. If there are pipelines nearby, operators may be able to direct the test gas to a processing plant and this process is called "in-line testing". This is not a practice that is feasible for some exploratory wells as there may not be any pipelines and processing plants nearby.

Natural Gas Battery and Pipeline Flaring: This type of flaring can occur at producing field facilities such as wells, dehydrators, compressors and gathering pipelines. Flares burn off gas during emergencies, maintenance shutdowns, equipment failures and other upset conditions."

FMS Managing Energy Services (www.flaring.ca)

Crew performing a well test flare on drilling rig #10 (Performance Services) near Grande Prairie.

LITTLE CHICAGO

Main Street, Little Chicago, May 1938. (Provincial Archives of Alberta, P1946)

Throughout Turner Valley's oil producing days came a steady stream of oilfield workers from all over Canada and the U.S. boom towns popped up throughout the area. As activity shifted to different areas in the Turner Valley oilfields, the workers would move their houses, referred to as "tar paper shacks" or "skid homes" to the new location.

Today, the only sign of the once booming Little Chicago is a small monument (left) and a pumpjack (see page 3) still drawing oil from this historic site.

Cairn marking the former site of Little Chicago.

LITTLE NEW YORK (LONGVIEW)

Little New York (Longview) Main Street, May 1938. (Province of Alberta Archives, P1932)

As a result of this "shack environment", many boom towns were formed complete with gambling casinos and cat houses! Some of the colourful names attached to these communities were: Whiskey Row, Poverty Flats, Dogtown, Snob Hill, Cuffling Flats, Naptha, Mercury, Little New York and Little Chicago. By the 1950s when oil drilling activity moved to the Leduc area, the shacks disappeared.

Town of Turner Valley – www.turnervalley.ca

Motorcycles approaching Longview (Little New York) on Highway 22.

LIFE IN THE PATCH

Throughout Turner Valley's oil producing days workers came from as far away as the United Kingdom and the United States. The population multiplied as Turner Valley continued to grow and on February 25, 1930 it was incorporated as a village. Although small in size, it had typical boomtown forms of entertainment. With roughnecks enduring high risk and working long hours, they became prone to wild and excessive forms of entertainment.

The Brown Family in 1914. (left to right) Josephine, Robert, Mary, Joseph and Charlie. Joseph Brown worked on Dingman #1 and #2 (Calgary Petroleum Products #1 and #2).
(Glenbow Archives NA-5535-9, copied from PA-3296)

As the twentieth century progressed and the oil fields were depleted, wells were shut down and transient oil field workers moved on in search of their next rig. Family activities started to emerge with typical recreational pursuits like baseball, hockey, curling, golf, the hunt club and family get-togethers gradually became the norm. Both children of ranchers and oil workers attended local schools and joined local scouting and guiding groups. And oil continues to be produced from many of these fields today.

Oil worker playing guitar outside bunkhouse, Turner Valley, ca. 1938. (Glenbow Archives na-4614-25)

This residential street in Devon (above), the town that Imperial Oil built from scratch following the Leduc #1 discovery, reflects housing in the oil patch today.

Kids and their dogs in the yard of oil workers' housing in Turner Valley, ca. 1930. (Glenbow Archives na-2895-11)

Oil worker constructing home, Little Chicago (Royalties), Turner Valley, ca. 1938. (Glenbow Archives na-2895-13)

27

Offices and living quarters at British Petroleum's oil well #3, Wainwright field, Alberta. ca. 1924.

Skid home awaiting restoration at the Leduc #1 Energy Discovery Centre, Devon.

Offices and living quarters on Nabors drill rig #88 near Innisfail today.

Human society has passed through two huge and lasting changes which deserve the name revolution. The first, the Neolithic Revolution, began in 8000 BC and continued through thousands of years. Its effect was to settle people on the land, turning the hunter-gather into the farmer and making peasant agriculture the standard everyday activity of the human species. The second, the Industrial Revolution, gathered momentum in the 19th century just as Europeans were flocking to the prairies of Western Canada to escape the feudal conditions in Europe and to pursue their own dreams of land ownership through the government's Homestead Act – much to the consternation of the indigenous people who had lived here for thousands of years.

Alberta's economy in the late 19th century was, like in the rest of Canada's, agriculture-based. But in just a few decades, the arrival of fossil fuels and the internal combustion engine would revolutionize the world's economies – perhaps nowhere more profoundly than in the province of Alberta.

Throughout the 20th, century Albertans moved away from farms for the added prosperity of the city and millions of them raised their families on money earned in the oil patch. Imagine what the first family to homestead the piece of land in the photograph below would have thought had they known what untold wealth lay deep beneath their crops and homes.

The names of the three generations we see here have been lost in the mists of history but we do know they are pictured here in Devon in 1947 and the men worked for the Rigid Drilling Company. I love this photograph and those beautiful, expressive faces and would be delighted to learn who they are. John McQuarrie

Province of Alberta Archives P 1400

Abandoned homestead near Cold Lake.

WIND ENERGY CURRENTLY POWERS OVER ONE MILLION CANADIAN HOMES.

RENEWABLE ENERGY

An infinite source of clean power: Canada's bountiful resource. Wind is powered by the sun. In fact, all renewable energy, and even energy in fossil fuels, ultimately comes from the sun. The sun heats our planet to different temperatures in different places and at different times. This unequal distribution of heat is what creates wind as warm air rises and cooler air descends to fill the void. Wind is the ongoing movement of this air.

As the sun warms the earth, it in turn, warms the air above it, making it less dense or lighter. As the light air rises, it creates a low pressure zone near the ground. Air from surrounding cooler areas rushes in to balance the pressure. These are called local winds. Temperature differences between the polar caps and equator, as well as the rotation of the earth, produce similar results on a global scale, called prevailing winds.

So how much wind do we have in Canada? We have more than we could ever use, and it's free. Our vast landscape, our three windy coastlines, the plains and mountains all contribute to this endless resource. Canada has still only scratched the surface of its massive wind energy potential, which currently powers over one million Canadian homes. Tomorrow we hope to do even more. Denmark already gets over 20 per cent of its electricity from wind. If we did the same in Canada, we would have enough wind energy to power 17 million homes! As long as the wind continues to blow, there is a great future in wind energy.

Canadian Wind Energy Association – http://www.canwea.ca/wind-energy/index_e.php

Wind Park near Fort McLeod.

WIND

Modern wind turbines, which convert wind into electricity, sit at the top of towers. The wind spins the blades of the turbine to create mechanical power. This mechanical power is used to turn a generator and produce electricity. Cables carry this electrical current to transmission lines. From there, the electricity is transmitted, connecting the power produced in one area of Alberta to customers in other areas of the province.

Alberta currently has over 500 megawatts (MW) connected to the grid. That is four per cent of installed capacity, which is enough capacity to serve over 500,000 homes. Currently, Alberta is one of the leading provinces in the development of wind power. There is also more than 11,000 MW of wind generation projects that have applied for connection to the transmission system.

Unlike conventional generation (from sources like coal, natural gas or hydro), wind can stop or start blowing without notice.

Wind Park west of Pincher Creek.

The amount of power being supplied to the electric system must be balanced with the demand across the province at all times. Wind can change quickly; there is a risk of wind power causing an imbalance of the electricity system, which could possibly interrupt service. When wind suddenly stops, conventional generation must be immediately dispatched. When the wind suddenly starts, power must be exchanged with other provinces to offset the imbalance.

In April 2006, the Alberta Electric System Operator (AESO) introduced a 900 MW threshold. This "cap" was a short-term and temporary way to ensure system reliability until work could be done with electricity producers on ways to accommodate more wind. The "cap" has been removed. The AESO is currently setting up a framework to allow more wind to be added to Alberta's power supply. The AESO is responsible for making sure Alberta's reliable grid operation will be maintained.

Canadian Wind Energy Association – http://www.energy.gov.ab.ca/Electricity/pdfs/FactSheet_Wind_Power.pdf

This farmer near High River is producing wind power and grain with the potential to become ethanol.

SOLAR

Energy from the sun travels to the earth in the form of electromagnetic radiation. At any given time the amount of solar energy depends upon the weather, location and the time of year. The amount of solar energy that can be converted to useable energy is dependent upon the technology and the application used.

Solar and other renewable energy systems can be stand-alone, thereby not requiring connection to a power or natural gas grid. The sun also provides a virtually unlimited supply of solar energy. The use of solar energy displaces conventional energy which usually results in a proportional decrease in greenhouse gas emissions.

The operating costs of solar energy systems are much lower than the cost to operate a combustion furnace with an air conditioning unit. However, the cost to install a complete solar system can be higher than the cost to install a furnace and air conditioning unit.

In the Northern Hemisphere the south side of a building always receives the most sunlight, therefore buildings designed for solar heating usually have large, south-facing windows. Materials that absorb and store the sun's heat can be built into the sunlit floors and walls. The floors and walls will then heat up during the day and slowly release heat at night, when the heat is needed most.

Other solar heating design features include sunspaces and trombe walls. A sunspace (which is much like a greenhouse) is built on the south side of a building. As sunlight passes through glass or other glazing, it warms the sunspace. Proper ventilation allows the heat to circulate within the building. A trombe wall is a very thick south-facing wall which is painted black and made of a material that absorbs a large amount of heat. A pane of glass or plastic glazing installed a few inches in front of the wall helps hold in the heat. The wall heats up slowly during the day, then as it cools gradually during the night it gives off its heat inside the building.

There are many design features that help keep solar buildings cool in the summer; for instance, overhangs can be designed to shade windows when the sun is high in the summer, sunspaces can be closed off from the rest of the building, and a building can be designed to use fresh-air ventilation in the summer.

The sun can be used to heat water with solar water heating systems. Most solar water heating systems for buildings have two main parts: a solar collector and a storage tank.

Oil companies are using solar energy to help drive their pipeline pumps.

Solar panels adorn rooftops of housing in Ocotoks.

The most common collector is called a flat-plate collector. Mounted on the roof, it consists of a thin flat rectangular box with a transparent cover that faces the sun. Small tubes run through the box and carry the fluid – either water or other fluid, such as an antifreeze solution – to be heated. The tubes are attached to an absorber plate which is painted black to absorb the heat. As heat builds up in the collector, it heats the fluid passing through the tubes. The storage tank then holds the hot liquid. It can be a modified water heater, but is usually larger and very well insulated. Systems that use fluids such as antifreeze can be used to heat the water in a modified water heater. The hot antifreeze passes through a coil of tubing in the tank transferring the heat to the water.

Photovoltaic (PV) cells convert sunlight directly into electricity. PV cells are the solar cells that are often used to power calculators and watches. PV cells are made of semiconducting materials similar to those used in computer chips. When these materials absorb sunlight, the solar energy knocks electrons loose from their atoms, allowing the electrons to flow through the material to produce electricity. This process of converting light (photons) to electricity (voltage) is called the photovoltaic effect.

PV cells are typically combined into modules that hold about 40 cells. Approximately 10 of these modules can be mounted in PV arrays that measure up to several metres on a side. About 10 to 20 PV arrays can provide enough power for a household. For large electric utilities or industrial applications, hundreds of arrays can be interconnected to form a single large PV system. www.agric.gov.ab.ca

Sunlight flaring through early-morning mist .

Agriculture and oil have sustained millions of Albertans for over 100 years.
ca.1949   (Canada Science and Technology Museum, X31966)

Wheat has always been a basic component of Alberta's economy and, together with beef, was for much of the province's history its prime source of wealth. With the explosive growth of the oil industry agriculture has slipped into second place but there is a lot of cross-pollination between the two. A maze of pipelines run just below the surface of all those fields of grain and cattle and now the actual wheat itself is finding its way into petroleum products.

Sequence of colour photos at right reflects straight combining and harvesting wheat with Eldon Bara piloting the combine and Greg Sereda at the wheel of the truck.

AGRICULTURE AND OIL HAVE SUSTAINED MILLIONS OF ALBERTANS FOR OVER 100 YEARS.

BIOFUELS

Commercial production of transportation biofuels is in its infancy in Alberta. It has some exciting potential benefits for rural economies and the environment, as well as some uncertainties.

Ethanol and biodiesel are the two main types of transportation biofuels. Ethanol is produced from carbohydrates – such as starch from grain or sugar from sugar cane – using a fermentation-distillation process. On the prairies, wheat is the main feedstock. Grain-based ethanol production yields a by-product called distillers grain, which can be used as a high-protein feed. Biodiesel is made from vegetable oils, like canola oil or animal fats. It is produced by mixing the fat or oil with methanol and a catalyst.

Both biofuels are used as additives: ethanol is blended with gasoline and biodiesel with petroleum diesel. In comparison to the straight petroleum products, the blends provide advantages like reduced tailpipe emissions, lower greenhouse gas emissions and improved engine performance, as well as new market opportunities for the agricultural feedstocks.

"These two products are added to transportation fuels in relatively small percentages," explains Alan Ford of Alberta Agriculture and Food. "For ethanol, blends up to E10 (10 per cent ethanol) are quite common and require no engine modifications. Blends as low as B2 (2 per cent biodiesel) have some beneficial performance improvements. For blends higher than about B20, you have to be aware of how the engine will operate in cold conditions." Biodiesel use doesn't require engine modifications.

Capital costs for ethanol plants are fairly high; for example, a plant with a 150-million-litre annual capacity would cost about $130 million. The only commercial ethanol plant in Alberta is Permolex International's facility in Red Deer, which also produces bakery flour, vital wheat gluten, fuel grade ethanol and livestock feed. It also has the potential to produce higher protein vital wheat gluten as well as other grades of ethanol, and the remaining by-products can be further processed to produce incremental revenue streams. The facility has incorporated a co-generation plant to produce the electricity and steam required in the various processes. The plant's current ethanol capacity is about 28 million litres per year. Construction is underway to expand that to about 40 million litres.

At present there are two biodiesel plants operating in Alberta: Western Biodiesel Inc. at Aldersyde (19 million litres/year) and Kyoto Fuels Corporation (66 million litres/year) located in Lethbridge. The relatively low cost for small-scale biodiesel production is sparking interest in farm-level and community-based production. Because ethanol and biodiesel can be made from agricultural feedstocks, their production could lead to rural development opportunities. "We've got the capacity in terms of feedstock supplies to produce these products. The thing that is missing is the market demand that would pull it into the transportation fuel market," says Ford.

To create that demand, the federal government has proposed enacting a standard that would require Canada's transportation fuels to have a 5 per cent renewable fuel content. Many jurisdictions, including the United States, already use this type of requirement to encourage biofuel production and use.

If a national standard is put in place, the next challenge would be satisfying that demand, explains Ford. For example, about 5.5 billion litres of gasoline are used annually in Alberta, so a 5 per cent standard would involve about 275 million litres of ethanol just for Alberta. He says, "It's very difficult to start at relatively small volumes and work with the petroleum retailers that deal in hundreds of millions of litres." The federal government already has various policies and programs to encourage biofuel production and use, as have many provinces, including all three Prairie provinces.

Manitoba's Biofuels Act includes measures to encourage community participation in production and use of Manitoba-grown feedstock. Saskatchewan has a standard that currently requires 1 per cent ethanol in gasoline; the standard will rise to 7.5 per cent as soon as there's enough processing capacity in the province to satisfy it. Saskatchewan also has an incentive program for ethanol that is produced and consumed in Saskatchewan, as well as measures to encourage smaller-scale ethanol production.

Alberta's Renewable Fuel Standard (RFS) was announced as part of the Provincial Energy Strategy on December 22, 2008. As of April 2011, the Alberta RFS will require an average annual blend of two per cent renewable diesel in diesel fuel and five per cent ethanol in gasoline sold in Alberta. Qualifying renewable fuels must demonstrate at least 25 per cent fewer greenhouse gas emissions than the replaced fossil fuel.

Another key factor in the growth of transportation biofuel production is the price of petroleum. Ford says, "If oil prices stabilize at $60/barrel or go higher, then I think there will be more and more interest in biodiesel and ethanol because we could produce those products at a price that would substitute for some of the higher cost petroleum product."

Other factors that could influence biodiesel and ethanol production in Alberta include distance from the major Canadian markets in Ontario and British Columbia, competition with the rapidly growing U.S. biofuel industry, changing feedstock prices, changing by-product markets and changing biofuel technologies.

www.agric.gov.ab.ca

Husky Energy, Lloydminster upgrader.

Meridian Cogeneration Power Plant, a joint venture between Husky and Trans Alta, provides excess electricity to Sask Power.

Husky is Western Canada's largest producer of ethanol and was a pioneer in the marketing of ethanol-blended fuels. Husky opened a 130-million-litres-per-year ethanol plant at Lloydminster, Saskatchewan in 2006, and a year later completed a 130-million-litres-per-year ethanol plant at Minnedosa, Manitoba to produce ethanol for fuel and industrial use. Husky and Mohawk retail outlets market ethanol-blended gasoline under the Mother Nature's Fuel brand.

Ethanol, also known as ethyl alcohol or grain alcohol, is made from wheat, corn or sugar. Husky's ethanol is made by converting the carbohydrate (starch) portion of grain into sugar. The sugar is then converted to ethanol in a fermentation process.

The primary feedstock for Husky's ethanol facilities is non-food, animal-feed-grade wheat purchased from local growers. The Lloydminster and Minnedosa ethanol plants are each designed to take in 350,000 tonnes of grain per year. Husky is one of Western Canada's largest grain buyers, purchasing a total of 700,000 tonnes per year.

The process of turning grain into ethanol involves milling, cooking, fermenting, distilling and dehydrating. Grain is milled, mixed with water to produce a slurry, and then cooked. Enzymes are added to the mixture to convert starch into fermentable sugars. Yeast is added to start the fermentation process. The fermented mixture requires distillation to separate the ethanol from the solids and water.

Distillation requires heating the fermented mixture until it sends off a vapour. The ethanol and water vapour from the top of the distillation column is captured, cooled and condensed to a liquid. The liquid then passes through a dehydration system where the remaining water is removed via a molecular sieve. The resulting product is pure ethanol. The produced ethanol is treated or "denatured" by adding a small amount of gasoline to turn it into fuel-grade ethanol. The ethanol is then blended with gasoline at points in the distribution system or at blender pumps at Husky and Mohawk retail outlets. It is also sold to third parties for offsite blending. The heavy solids or mash that remain are dried to produce dried distillers grain with solubles (DDGS), a high-protein feed supplement sold to livestock producers.

Ethanol-blended fuels increase engine performance by:
- Reducing those pre-ignition, "knocking" or "pinging" noises which under severe conditions can result in serious engine damage.
- Keeping fuel tanks and systems clean of contaminants and water through its natural solvent and water-absorbing properties.
- Acting as a natural gasoline-line antifreeze in winter. Adding ethanol results in a higher octane rating. Husky creates 89 to 94 octane Mother Nature's Fuel by blending up to 10 percent ethanol with regular or premium gasoline. The higher octane levels of cleaner-burning ethanol-blended fuels are beneficial to high-performance engines.

www.huskyenergy.com

Truck-load of wheat arriving at the refinery. The Company purchases 350,000 tonnes of wheat per year from local producers for the manufacturing of ethanol at Lloydminster.

Alberta has an exciting opportunity to produce energy from waste. The waste can be either manure or food-processing wastes. The province has a significant number of large agricultural operations which produce a considerable amount of organic waste in the form of manure, crops, crop residues and animal remains. Handling such large amounts of organic waste, especially manure, in an environmentally friendly manner is a challenge.

Producers, scientists and other stakeholders are exploring various options to tackle this issue, and using anaerobic digesters (AD) is a promising one among them. Anaerobic digesters are specially designed tanks used to facilitate the anaerobic digestion process under a controlled atmosphere. Anaerobic digestion is a natural process that occurs in the absence of air. During this process, micro-organisms stabilize the waste organic matter and release biogas as a by-product.

Biogas consists mainly of methane and carbon dioxide gases. Burning biogas can produce energy like natural gas. The energy produced using biogas is renewable, unlike natural gas. Some scientists and academics anticipate that renewable energy sources will be preferred over the natural fossil fuel energy sources in the

Stabilized organic wastes from a digester, known as digestate, contain less odour than the unstable waste or no odour at all, yet retain almost all the nutrients from the feed material. Applying the digestate to cropland may replace commercial fertilizers, so anaerobic digesters can bring several benefits.

The main advantage of anaerobic digester technology is that it produces renewable energy while stabilizing waste organic matter. This renewable energy can be a part of solving some issues such as climate change and high energy costs. This system reduces odour and the risk of ground water contamination originating from intensive livestock operations. Adopting this technology may also increase employment opportunities in rural populations, as it requires trained operators.

Despite these advantages, this technology remains expensive and unproven in terms of substantial economic benefits.

Government of Alberta, Agriculture and Rural Development

Mother cows and calves on range land near Longview Alberta. Typically, calves will be weened in the fall and shipped to feedlots like the one adjacent to the Hairy Hills site near Vegreville which is described at right.

Integrated bioMass Utilization System (IMUS™) is a patented system to derive renewable energy from high-solids, high-fibre, organic wastes such as manure from outdoor feedlots, food industry residues, and municipal wastes. The IMUS™ system that the Integrated bioRefinery™ will use has been in place and operational for over two years at the Hairy Hills site near Vegreville, and will be expanded approximately fourfold to match the needs of the Integrated bioRefinery™ scheduled to go on line in 2011..

Biogas is a renewable energy product similar in many ways to natural gas; so similar, in fact, that it can properly be referred to as renewable natural gas! It is produced by trillions of tiny bacteria-like organisms in the IMUS™ anaerobic digesters, effectively making IMUS™ a natural gas well! But better than a regular gas well, IMUS™ never runs out of renewable biogas!

Raw biogas is composed of a combination of several gases: methane (the main component of the natural gas that comes to your home), carbon dioxide, water vapour, and traces of some other gases. In IMUS™, we clean the biogas so that it contains only high proportions of methane and some carbon dioxide; we call this "polished biogas!"

Further processing of polished biogas to make it chemically the same as the natural gas used to heat homes worldwide is possible too; the resulting product is often referred to as renewable natural gas.

The Growing Power Hairy Hill Intregrated Integrated bioRefinery™ will make use of polished biogas to produce green electricity, steam, and hot water in our bioUtility™ that will fuel the workings of the Integrated bioRefinery™.

The bioUtility™ is the heart of the Integrated bioRefinery™, producing green-electricity, steam and hot water from biogas, that are in turn used in the IMUS™ and ethanol production plants.

BioUtilities are very useful for the appropriate coversion of bioGas into usable forms of energy, and our sister company Highmark Renewables Research can assist other biogas producers, such as sewage treatment plants, in maximizing their economic returns with a bioUtility™ unit.   www.growingpower.com

WE'RE NOT JUST GREEN, WE'RE GROWING POWER!

Growing Power Hairy Hill is developing the first Integrated bioRefinery™ in Canada.

Sunset on Nabors #88 and the work goes on.

THE RIGS

CABLE TOOL RIG

When the first commercial oil well in North America was drilled at Petrolia, Ontario in 1858, a cable tool rig was used. Cable tool rigs had been in use for hundreds of years. They were used to drill for fresh water or for brines that were evaporated for salt.

Cable tool rig floor. Leduc #1 Energy Discovery Centre, Devon, Alberta.

The cable tool rig was also used in such early Alberta fields as Turner Valley (below) where it was gradually phased out by the metal derrick. Cable tool drilling first involved the construction, on site, of an 82-foot high wooden derrick, built on a foundation of huge timbers. The legs were made of 2" x 12" rough lumber laminated to form a right angle similar to angle iron. These were braced with horizontal girts and diagonal struts. Interestingly, all of the girts and beams were designed by engineers and produced by carpenters and saw mills so each piece of wood fitted exactly in its place in the derrick. The pieces were then catalogued and listed for sale by oilfield supply companies. Any time a piece of the derrick was damaged, the supply store could replace that piece from stock.

The well is drilled by raising and lowering a heavy bit and slowly pounding a hole into the ground. After some drilling, the bottom of the well became clogged with rock chips. The bit was then raised and a bailer lowered into the well to scoop out the rock chips. After the bailer was removed, the bit was then lowered into the well to pound it deeper. This sequence was repeated about every three feet.

Cable tool rig in foreground with metal derricks in background. Turner Valley, ca. 1935.

Canadian Museum of Science & Technology 31826

Cable tool rig in foreground with jack-nife derrick in background. Leduc #1 Energy Discovery Centre, Devon.

ROTARY RIGS

A drilling rig drills a hole in the ground or the ocean floor to a predetermined depth for optimum production. Once the hole is complete, the drill crew runs casing into the well bore and pours concrete around the casing.

Rotary rigs drill the vast majority of wells today, including all medium and deep wells. The rotary rig consists of four major systems. These include the engines, and the hoisting, rotating and mud systems.

The engines supply the power to the rig. Most Alberta rigs use a single engine to power the draw-works and rotary table. These engines are diesel fuelled and are rated between 425-550 hp. The power is used primarily to turn the drill string and raise and lower equipment in the well.

Performance Services crew members Charles Patriquin and Aaron Armstrong working Service Rig #10 and dealing with a little "kick" (right) near Grande Prairie, Alberta.

Oil well blowing probably near Turner Valley, Alberta. (Glenbow Archives NA-2736-2)

Engines also supply the electricity used on and around the rig. Electrical power is supplied, usually, through two generator sets. The rig can run with one of these units but it would run at close to maximum output at night. The second provides for back-up and allows for other options.

Rigs also employ one or two engines to power the mud pump. There's more to drilling than simply rotating the bit. Mud is circulated while the drilling proceeds. Powerful pumps move the fluid down the pipe, through the bit and back to the surface, carrying the cuttings and other debris with it. Thus, on a rotary rig (unlike the cable tool), drilling can be continual as stopping to bail the cuttings is no longer required. The drilling mud also stabilizes the walls of the hole. The hoisting system is used to raise and lower and to suspend equipment in the well while the rotating system is used to cut the hole.

The bit screws into the bottom of the drill collars. The most common bit is the tricone bit, which has three rotating cones. The cones have teeth that are designed to chip and flake away the rock as the bit is rotated. Some tricone bits have hard tungsten carbide buttons instead of teeth. Another type of bit is the diamond or button bit, which has diamonds embedded in the bottom and sides. Different bits are used for different hardness formations.

Bits wear out after eight to 200 hours of rotation, with an average bit wearing out after about 24 hours. A worn bit can be detected by the noise on the derrick floor that the rotating drill pipe makes and by a decrease in rate of drill penetration. "Making a trip" is necessary for changing the bit. All the pipe is pulled out of the hole (tripping out) and stacked in the derrick. The bit is then changed and the pipe put back into the hole (tripping in). This takes rig time and costs money. The deeper the well, the longer the trip takes.

Adapted from: www.lloydminsterheavyoil.com

Drilling platform floor detail. Leduc #1 Energy Discovery Centre, Devon.

Working on drilling rigs is a physically demanding job, with long hours, hard work, extreme weather conditions and long shifts. Drilling rigs normally run twenty-four hours a day – in all kinds of weather – until the well is finished. The crews that run the rigs usually work twelve-hour shifts. Time off varies from company to company and from rig to rig but the most common work schedule is twelve hours on and twelve off for fourteen days straight and then seven days off. The two-week work period will be split between day and night shifts. Leasehands and floorhands work outdoors, often year round in remote locations, exposed to extremes in weather as well as to the dirt, dust, noise and fumes common around rigs. The work is physically demanding and involves working on wet, slippery rig floors, near or with heavy tools and moving machinery, and then there is exposure to chemical substances such as paint, motor oil and drilling mixture additives. Because the work is potentially hazardous, all drilling crew members must think and act quickly, always follow standard safety practices and participate in safety meetings and emergency procedure drills.

Working in the mud on a Performance servicing rig near Carsland.

(Opposite) Driller Mark Shore and roughneck Tyler Rook working the floor on Nabors rig #19 near High River.

COIL TUBING RIG

The quest to develop the ideal drilling system has led to the combination of coil tubing well servicing units with state-of-the-art programmable AC electric drilling rigs; the result is the latest generation of Hybrid Coil Tubing / Top Drive Drilling rigs. These rigs provide incredible flexibility and allow the optimum drilling technique to match the section of hole drilled; e.g., jointed pipe with the Top Drive for surface hole, coil tubing for the main hole. These rigs also offer improved ergonomics (reduced pipe handling and rig floor activities) which improves safety performance. They can be moved quickly and have a small surface footprint for reduced environmental impact.

Sunrise finds the night crew hard at work on Nabors #88 near Claresholm.

Coil tubing drilling is an alternative technology to conventional jointed pipe drilling. Coil tubing is a continuous reel of steel tube. Jointed pipe are individual lengths of steel pipe which must be screwed together. Coil tubing drilling has an intrinsic technical advantage; because there are no connections; drilling speed (rate-of-penetration) and trip speeds (running the drill string in and out of the well) are greatly increased.

There are three focus applications where, currently, coil tubing drilling offers the most obvious advantages; shallow wells, re-entry wells, and underbalanced wells.

Coil tubing drilling of shallow wells in soft formations has demonstrated a drilling speed two times faster than with jointed pipe and trip speeds three times faster than jointed pipe. This advantage is reduced when drilling deeper, harder formations. The small diameter of coil tubing makes it ideally suited for re-entry drilling where you are drilling from the inside of steel casing already cemented in the well. Coil tubing directional drilling systems are available that can deliver the horizontal wellbore trajectories to re-access or extend the wellbore into productive formation.

A major challenge with underbalanced drilling has been the difficulty managing bottom hole pressure when making connections, both drilling and tripping. Coil tubing drilling provides significant advantages since there are no connections and the ability to continuously pump allows more flexibility in bottom hole pressure management.

Hybrid Coil Tubing / Top Drive Drilling Rigs incorporating state-of-the-art programmable AC electric controls provide a flexible drilling system that reduces time, reduces costs, increases safety and reduces environmental impacts.

www.nabors.com

Beautiful replica of an early cable tool drilling rig at the Leduc #1 Energy Discovery Centre in Devon, Alberta.

Below you are inside the control room of Nabors Rig #88, a modern, coil tubing drill at work near Innisfail with Pat Christiansen working the panel . At right, Tobey Hofer can be seen on the same rig loading pipe. These photographs also reflect a hundred years of progress in the evolution of oil drilling technology.

DIRECTIONAL DRILLING

Because of dipping beds of hard and soft rocks, drillers used to have a hard time keeping a well going straight down. If the bit hit a subsurface hard rock layer with a dip greater than 45 degrees, the bit would tend to be deflected. Drilling deviations caused by dip deflection can be prevented when drilling in areas where the geologic structure is known. By reducing the bit pressure and allowing the bit to do the work, deviation can be eliminated.

Today, drilling contracts often have a clause stipulating that the well deviate no more than a few degrees from vertical. Modern rotary rigs can be controlled so that the well is drilled at a predetermined angle (directional or deviation drilling) and ends up in a predetermined location.

In the past, some wells were "accidentally" drilled to drain oil out from under adjacent leases. There are also many legitimate reasons for drilling a crooked hole. If a well is on fire and cannot be approached, a relief well can be drilled at a safe distance from the wild well. Heavy drilling mud is then pumped from the relief well through the subsurface rocks and into the wild well to control it. It is more economical to drill a crooked hole to test several potential petroleum reservoirs than to drill several wells to each reservoir.

More recently, directional drilling and slant drilling have become common in Canada. Slant drilling allows more than one (usually 4, 6 or 8) wells to be drilled off the same lease site (pad). Directional drilling (where the direction of the well changes from time to time) is usually associated with some enhanced recovery technique (EOR), such as horizontal wells being drilled for SAGD (steam assisted gravity drainage).

Performance Service rig #2, Carseland, Alberta.

Joseph Brown, assistant driller, Dingman #1 well (Calgary Petroleum Products #1), Turner Valley, ca. 1914. (Glenbow Archives NA-5262-39)

TOOL PUSH

Dave Dylke, Tool Push on Nabors Rig #51 surrounded by canola crop near High River.

Crew at Dingman #1 well (Calgary Petroleum Products #1) Turner Valley, ca. 1914-1917. Glenbow Archives, NA-5262-44)

Floor Hand/Roughneck (two to a crew) works on the drilling floor when tripping: pulling and running pipe, operating the pipe tongs and elevators. When not tripping he takes cutting samples and does general rig maintenance.

Derrick Hand works up the derrick when pulling drill pipe out of the hole (tripping pipe). His job is to latch and unlatch the elevators and organize the pipe in the derrick. When not tripping, he is responsible to ensure that the drilling mud is of the proper specifications.

Driller is the individual who runs the draw-works, which is the hoisting unit to pull and run pipe and provides power for the rotary table. He is the shift supervisor of the drilling crew. He has normally spent many years on a drilling rig gaining the knowledge to become a driller. There is one driller on each shift.

Mud Man is responsible for setting up and maintaining a drilling mud program that is suitable for the particular drilling conditions expected to be encountered and to maintain well control while drilling.

Rig Manager/Toolpush has normally had a long career in the drilling business and is very experienced in all types of drilling operations. Acts as the rig manager, on call 24 hours a day and is the drilling company's representative on the drilling site. The term "tool pusher" came from cable tool rig days when he instructed the tool dresser on creating the drilling bit in a forge on site.

Consultant: While the toolpusher represents the drilling rig owner, the consultant looks to the interests of the oil company that has hired the drilling company for a given job.

Crew of Nabors rig #19 pose for group photo near Claresholm. (Left to right, Mark Shore, Tyler Rook, Rocky Randall, Brandon 'Ox' Erdos and Morgan Wholers.)

Workers at Viking #2 well, Viking, Alberta. (Glenbow Archives NA-1072-17

Drilling crew on Nabors drill rig #88, Innisfail, Alberta.
(Left to right, Wade Trowbridge, Josh Beals, Kyler Cleverly, Chase Sommerfeldt, John Rogers and Shane Trowbridge.

When most people hear the term "oil rig", they typically think of a drilling rig operation that drills new wells and installs casing pipe. Service rigs are different from drilling rigs in many aspects; for starters they do different things to a well. While drilling rigs are used to drill a new well, service rigs perform services on wells that have already been made. Service rigs perform such tasks as completing wells that have just been drilled, fixing wells that do not produce oil and abandoning old wells that have stopped producing.

Workers on Performance Drilling Rig #2 at Carseland, Alberta. (All colour photos this spread).

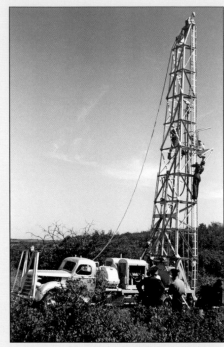

Erecting oil derrick on well site, Kamsack, Saskatchewan, 1945. (Glenbow Archives NB-43-2)

Occasionally, oilfield work can be a little muddy.

A jackknife rig (photos this spread) is a steel framed rotary drilling derrick that is designed so it can be stood up and laid down in one piece. This design replaced the conventional, metal derrick that had to be assembled and disassembled on site, piece by piece.

Transporting oil derrick to well site, Kamsack, Saskatchewan. 1945 (Glenbow Archives NB-43-1)

Performance Services rig #2 being readied for a day's work near Carseland.

In 1987, Nabors' scope of business was limited to contract land-drilling services on the North Slope of Alaska, Western Canada and in the Rocky Mountains region of the United States. Since that time Nabors has expanded its contract land-drilling services into most major oil and gas producing regions of the world. In addition, Nabors has broadened its oil and gas contracting services to include offshore drilling and workover and US land well-servicing and workover.

More recently, Nabors has been actively expanding its business into areas that supplement its primary business as a drilling contractor. These businesses provide their customers with information technology, drilling systems, engineering, transportation, construction, maintenance, well-logging and other support services in selected domestic and international markets.

Nabors is the largest drilling contractor in the world, conducting oil, gas and geothermal land drilling operations in the U.S. lower 48 states, Alaska, Canada and internationally in more than 30 additional countries.

Nabors' wholly-owned subsidiary, Canrig Drilling Technology, Ltd. manufactures, markets and services a full range of portable and fixed configuration top-drive systems for installation on most land and offshore drilling rigs.   www.nabors.com

Construction of a large 'triple' rig nearing completion in the yard of Nabors Drilling in Calgary. Very soon it will be added to the fleet and see service in places like the one pictured at right near High River, Alberta.

Nabors rig #19 near High River.

National Oilwell Varco designs, builds and supports the widest range of self-propelled and trailer-mounted well servicing, work-over and drilling mobile rigs in the world. The Canadian operation in Nisku, Alberta specializes in these large, mobile rigs designed for drilling in Kuwait. When completed, each of these rigs will be disassembled and loaded aboard 20 large tractor-trailer rigs where they will begin their long journey to Kuwait. At just under 20 million dollars a copy, they are one of the most expensive, land-based drilling rigs in the world. www.nov.com

To get an idea of the size of these rigs, find the worker on the drilling floor of the photo at left. Another interesting feature of these rigs is that the blue portion at the top folds down to allow the drills to pass under the high power lines that criss-cross Kuwait without requiring any disassembly of the derrick.

Two giant drill rigs nearing completion. Tires weigh 8000 kgs. each and are almost four metres in diameter.

At just under **20** million dollars a copy, they are one of the most expensive, land-based drilling rigs in the world.

Workers adding finishing touches to two rigs in yard at Nisku, Alberta. (Inset above left) Rig on location in Kuwait. (courtesy National Oilwell Varco)

A BARREL OF OIL

C.C. Snowdon plant, Calgary, Alberta. Interior of refinery, showing barrels of oil and employees. Located at 1810-1840 11th Street E, Calgary. ca 1914 (Glenbow Archives NA-2726-8)

In the age of metric, the ever-popular "barrel" remains the world standard of oil measurement. For comparison sake, it converts to either 35 Imperial Gallons or 42 U.S. gallons. For those metrically inclined, one barrel equals 158.987 litres.
One barrel of oil is the same as:

- 159 litres (about 80 large fizzy drink bottles)
- 35 gallons (enough to fit in the petrol tanks of about 4 cars)
- 280 pints (a lot of bottles of milk)

Crude oil prices behave much as any other commodity with wide price swings in times of shortage or oversupply. The crude oil price cycle may extend over several years, responding to changes in demand as well as to OPEC and non-OPEC supply.

Hauling barrels of oil to Calgary from Turner Valley, ca. 1914. (Glenbow Archives PA-3689-1)

Long gone are the days when actual barrels were used. Here Rick Harper is delivering crude oil to the Hartell facility where it will enter the pipeline system enroute to a refinery

(Lane & Mitchel photo, Glenbow Archives VA-953-10)

Filling a barrel at Dingman #1 well (Calgary Petroleum Products #1), Turner Valley, Alberta, 1914. Joseph Brown, an American driller recruited by Archie Dingman to build and run Dingman #1 is the one pouring the oil on the left. He remained in Black Diamond with his family through the first two years of Canada's first oil boom.

Truck loading station, Suncor refinery, Edmonton.

Tank cars at Imperial Oil refinery, Calgary, 1943.
(Glenbow Archives NA-4281-28)

Imperial Oil Company tank trucks fuelling General Motors Diesel Engine on a demonstration trip to showcase its new diesel engines. (Glenbow Archives IP-2b-59)

Pumping station is the only clue that there is a pipeline buried beneath this pastoral scene just a few kilometres south-east of Calgary.

Lowering completed section of pipeline into trench, ca. 1955. (Canada Science and Technology Museum, CN003655)

MANY PEOPLE ARE SURPRISED TO DISCOVER THAT THERE ARE ALMOST 400,000 KILOMETRES (250,000 MILES) OF ENERGY-RELATED PIPELINES IN ALBERTA.

Nearing completion, this pipeline will bring Fort McMurray bitumen 493 kilometres to Shell's Scotford upgrader north of Fort Saskatchewan.

History of Pipelines

Pipelines were developed to transport products to market, products – such as crude oil, natural gas, gasoline, aviation fuel and the raw materials for plastics, fertilizers and medicines – that are used by Canadians every day. Pipelines are important to Canadians because they are the safest and most efficient means of transporting the products that support our standard of living.

The first gathering systems in North America were constructed of hollow logs and were used to transport natural gas short distances from well sites to nearby towns. Distribution systems were also constructed to deliver the natural gas to buildings and street lights. In the early 1900s, there were only a few transmission pipelines in Canada. One ran from oil fields in Ohio to refineries in the Sarnia area. Another ran from Bow Island to Calgar and, at 270 kilometres (168 miles), was the longest pipeline in Canada at its completion in 1912.

During WWII, because of security concerns, oil pipelines were built from Portland, Maine to Montreal, and from Norman Wells, Northwest Territories to Whitehorse, Yukon. The latter, known as the Canol Pipeline, was only used for about a year, while the former is still in use.

Welding pipeline from Bow Island to Calgary, 1930. (Provincial Archives of Alberta, P1973)

Pipelines have gotten a little bigger since 1930 but the basics of moving oil and gas remain the same.

Welders putting finishing touches on 46" pipeline from Fort McMurray to the Shell upgrader near Edmonton.

Major discoveries in Alberta during the 40s and 50s, such as Leduc, Pembina, and Swan Hills, spurred construction of the Interprovincial Pipelines (now Enbridge Pipelines Inc.) crude oil pipeline stretching from Edmonton to Superior, Wisconsin in 1950, and on to Sarnia in 1953. The Trans Mountain Pipeline now owned and operated by Kinder Morgan Inc. was also completed in 1953 to transport crude oil from Edmonton to Vancouver.

The Westcoast Pipeline (now owned and operated by Spectra Energy Corp.) began transporting natural gas from north-east British Columbia to the B.C.-U.S. border in 1957.
    Construction of the TransCanada Pipeline began in 1957 to provide a secure source of natural gas for Central Canada. It was completed the following year. Also in 1957, TransCanada's Alberta System, referred to as NGTL or NOVA, began operation.

As demand grew, these pipelines were expanded and additional routes were opened to carry Canadian crude oil and natural gas to markets in California, the Pacific North-west, the Rocky Mountain States and the Midwest United States.

Welders preparing pipe to be joined.

Pipeline surfaces in canola field south-east of Calgary.

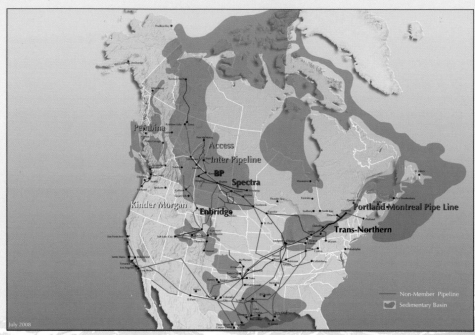

Liquid pipelines. Courtesy of Canadian Energy Pipeline Association

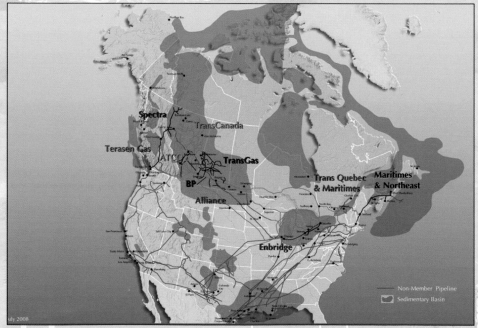

Natural gas pipelines. Courtesy of Canadian Energy Pipeline Association

Types of Pipelines

There are three major types of pipeline used to transport hydrocarbons, as defined by throughput: crude oil pipelines, natural gas pipelines and product pipelines.

Crude Oil Pipelines

Approximately 2.65 million barrels of crude per day travel through Canada's crude oil pipeline network which includes everything from small-diameter plastic gathering lines to steel conduits more than one metre in diameter.

Small-diameter (five to 15 centimetres) gathering system pipelines in individual fields carry oil from wellheads to a central facility in the field called a battery.

Larger lines (up to 20 centimetres in diameter connect groups of batteries with local refineries or with still larger trunk lines (up to 120 centimetres in diameter) which feed refineries across the country.

Where gathering systems are not available, oil is transported by truck to trunk lines. Crude oil and refined products are also transported by ship and by railway.

The oil is moved along the pipelines by powerful centrifugal pumps spaced along the line at intervals depending upon pipeline size, capacity and topography.

Different types of oil, heavy oil, bitumen and natural gas liquids travel in batches at between four and eight kilometres per hour. Because the different batches in a pipeline move as a continuum at the same speed, there is no need to separate them. Mixing only occurs where two batches come in contact with each other and these small volumes, known as transmix, are reprocessed.

There are currently 16 refineries in Canada: two in British Columbia, three in Alberta, one in Saskatchewan, four in Ontario, three in Quebec and three in the Atlantic Provinces, all of which are connected to the pipelines system.

Tanker truck in the distance is delivering crude oil to the Hartell facility south of Calgary where the crude will enter the pipeline system enroute to a refinery.

Natural Gas Pipelines

Approximately 484 million cubic metres (17.1 billion cubic feet) of natural gas per day travel through Canada's natural gas pipeline network which, like oil pipelines, comprises everything from small-diameter plastic gathering lines to steel conduits more than a metre in diameter.

Unlike crude oil, natural gas is generally delivered directly to the consumer by pipeline. However, it begins that journey in a manner similar to crude oil. Gas wells are connected to small-diameter (five centimetres to 15 centimetres) gathering systems that take the gas to a gas processing facility. Gas processing facilities, usually referred to as gas plants, vary in size from small compression facilities that are mounted on moveable platforms and that remove impurities and water from the gas, to large gas plants that also remove sulphur and carbon dioxide. Some gas plants also extract ethane, propane and butane which are referred to as natural gas liquids or NGLs. The generally dry gas may then be compressed prior to moving into the transmission system which consists of steel pipe from 50 centimetres to more than a metre in diameter.

Gas flows through the system from areas of high pressure to areas of low pressure through the use of compressors, turbines similar to jet engines that increase the pressure of the gas up to 10,300 kilopascals . Compressor stations are placed at regular intervals along the pipeline to increase line pressure which is reduced due to friction of the gas moving through the pipe.

Transmission line compressors are most often driven by gas turbines with the necessary fuel being taken from the pipeline. Where electricity is preferable, electric motors may be used to drive compressors.

Transmission systems move the gas across great distances to local distribution companies or gas utilities, where the pressure is reduced and the gas enters a distribution main for local delivery to service lines connected to individual homes or businesses.

Product pipelines carry refined products such as gasoline, diesel, jet fuel, heating oil and lubricants from refineries to terminals or local distribution centres. Like oil pipelines, centrifugal pumps move the products through the line. Pipe diameters range from 20 centimetres to as high as 106 centimetres.

Canadian Energy Pipeline Association, www.cepa.com

NEW PIPELINE PROJECTS

Northern Gateway

The proposed Enbridge Northern Gateway Project involves a new twin pipeline system running from near Edmonton, Alberta, to a new marine terminal in Kitimat, British Columbia. One pipeline will export petroleum while the second will import condensate, a by-product of natural gas used in transporting petroleum.

The pipelines will be buried at a depth of one metre in a 25-metre-wide right-of-way and will be 1,170 km in length.

From selecting the route to sourcing our materials, Enbridge is committed to providing the highest standards of pipeline safety and environmental care on Northern Gateway. The permanent right-of-way (ROW) is a strip of land approximately 25 metres wide containing the two pipelines. Additional temporary construction work space will be required along the pipeline right-of-way during construction.

Enbridge has strict environmental standards to ensure maximum protection when they build and operate pipelines. Because the land is rebuilt to exacting standards, right-of-ways can typically be used as they were prior to construction.

www.northerngateway.ca

Keystone

An innovative and cost-competitive solution to a growing North American demand for energy, the Keystone Pipeline System will link a reliable and stable source of Canadian crude oil with U.S. demand. Upon completion, the Keystone Pipeline System will be comprised of the 3,461 kilometre (2,151 mile) Keystone Pipeline and the proposed 3,134 kilometre (1,959 mile) Keystone Gulf Coast Expansion Project (Keystone XL). TransCanada affiliates will build and operate the Keystone Pipeline System.

The Keystone Pipeline will transport crude oil from Hardisty, Alberta to U.S. Midwest markets at Wood River and Patoka, Illinois and to Cushing, Oklahoma. The Canadian portion of the project involves the conversion of approximately 864 kilometres (537 miles) of existing Canadian Mainline pipeline facilities from natural gas to crude oil transmission service and construction of approximately 373 kilometres (232 miles) of pipeline, pump stations and terminal facilities at Hardisty, Alberta. The U.S. portion of the project includes construction of approximately 2,219 kilometres (1,379 miles) of pipeline and pump stations.

The Keystone Pipeline will have an initial nominal capacity of 435,000 barrels per day in late 2009 and will be expanded to a nominal capacity of 590,000 barrels per day in late 2010. Keystone has contracts with shippers totalling 495,000 barrels per day with an average term of 18 years.

www.transcanada.com/keystone/

Mackenzie Gas Project

The Mackenzie Gas Project proposes to build a 1,196-kilometre pipeline system along the Mackenzie Valley. It would link northern natural gas producing wells to southern markets. The main Mackenzie Valley Pipeline would connect to an existing natural gas pipeline system in north-western Alberta. Four gas producer companies, Shell, Imperial Oil Resources, Conoco Philips Canada and ExonMobile Canada are working with the Aboriginal Pipeline Group (APG) to make this happen.

The proposed Project crosses four Aboriginal regions in Canada's Northwest Territories (NWT): the Inuvialuit Settlement Region, the Gwich'in Settlement Area, the Sahtu Settlement Area and the Deh Cho Territory. A short segment will be in north-western Alberta near the NWT border.

The natural gas exploration and development companies involved in the Mackenzie Gas Project have interests in three discovered natural gas fields in the Mackenzie Delta, Taglu, Parsons Lake and Niglintgak. Together, they can supply about 800 million cubic feet per day of natural gas over the life of the project. Other companies exploring for natural gas in the north are also interested in using the pipeline. In total, as much as 1.2 billion cubic feet per day of natural gas could be available initially to move through the Mackenzie Valley Pipeline.

www.mackenziegasproject.com

Mackenzie Delta. Fritz Meuller Photography – www.fritzmueller.com

Crew performing a well test flare on service rig #10 (Performance Services) under the watchful eye of rig manager Brad Trentham (right) near Grande Prairie.

The new millennium brought a return to healthy oil and gas prices, and strong activity levels followed in step. But it also brought a redoubled push by industry regulators to shrink the oil and gas industry's environmental footprint and a multi-billion-dollar environmental services industry began to emerge.

The Alberta environment minister at the time, Gary Mar, put it this way, "There's a demand that our air is clean, our water is clean, and the fish in our rivers, streams and lakes are edible."

Alberta tightened its air emission rules the year before when public anger over wasteful flaring boiled over. The new rules called for a 70 per cent reduction in flaring over seven years. The remaining flares would have to burn harmful substances at near 100 per cent efficiency. Also, benzene emissions from glycol dehydrators were to be reduced by 90 per cent by 2007.

More stringent soil remediation targets were put in place for reclaiming exhausted wellsites and other facilities. The standard brought in a benchmark of 1,000 parts per million of hydrocarbons. Tighter topsoil replacement and revegetation rules resulted in one in five reclamation applications failing.

On Crown land, rules protecting wildlife habitat and preserving unique landscapes stalled some oil and gas development. Environmental concerns would go well beyond doing as little harm as possible to ensuring that cumulative effects of all developments were sustainable.

Accounting for human populations and their effects on the environment was added to the mix. New roads and seismic lines that also happened to create access for hunters, snowmobilers and off-road vehicles to formerly remote areas came under scrutiny.

Meanwhile, in the wind-swept foothills and plains of southern Alberta, Enmax, Enbridge, Canadian Hydro Developers, Vision Quest Windelectric, Suncor and a handful of other oil and gas companies were either hoisting up or planning wind farms to feed clean energy into Alberta's electrical grid.

Cogeneration projects at oil and gas facilities as well as biomass schemes were catching the headlines.

And then there was the looming Kyoto Accord greenhouse gas headache, which still awaited ratification by Canada. Canada's emissions were steadily rising, making it increasingly clear to the industry that only the most draconian efforts would bring them in line with Kyoto targets. *O ilweek*, June 2008 – 60th Anniversary Edition
www.oilweek.com

pumpjacks and oil field equipment on display at the Leduc #1 Energy Diccovery Centre, Devon.

LEDUC

"The history of Leduc can be traced back to 1889 when Robert Taylor Telford settled on a piece of land near a scenic lake. This piece of land would become the cornerstone of the new town. During those earlier years, Robert Telford was the first postmaster, first general merchant and first justice of the peace in the settlement that had become informally known as Telford. He also later served the community as Mayor and as a member of the Legislative Assembly.

In 1890, a government telegraph office was being set up by Mr. McKinely, a settler in the area. He needed a name for the place and said, "We shall name it after the first person who comes in." In through the door came Father Leduc.

In 1899, Lieutenant Governor Dewdney of the Northwest Territories, decreed that the settlement of Telford should be renamed "Leduc" in honor of the noted Roman Catholic missionary, Father Hippolyte Leduc, who had served the area since 1867 and later went on to become the Vicar General of the Diocese of Edmonton.

The municipality of Leduc was officially incorporated as the Village of Leduc on December 15, 1899, grew to attain town status on December 15, 1906 and eventually became the City of Leduc on September 1, 1983.

Leduc continued to grow and prosper as a major stopping point between Edmonton and Calgary. However it was not until February 13, 1947, when oil was first discovered at Leduc #1, that the new era was ushered in. This discovery was the beginning of a massive economic revolution for Leduc and Alberta. Alberta changed from a predominately rural and agricultural province to an urban economy dominated by the oil and gas industry.

With more than 20,000 residents, the City of Leduc is a thriving community that is strategically situated along Alberta's major road and rail connections and is ideally located near the Edmonton International Airport, connecting Leduc to the world."

www.city.leduc.ab.ca

View to the east along Main Street, toward Telford Lake, Leduc, 1899. (Provincial Archives of Alberta, A1819)

View to the east along today's 50th Avenue at 48th Steet, Leduc. Telford Lake is now obscured by trees.

The site of the original Leduc discovery well is actually located in the town of Devon, about 25 kilometres (15 miles) north-west of Leduc. But on that February afternoon in 1947 when the well came in, Devon did not exist, so it was named for the nearest town at the time, Leduc. Imperial Oil soon realized that the tide of rough and tumble oil workers were not mixing well with the quiet and reserved rural townspeople so, in typical oil-company style, they immediately began construction of the new town of Devon, just north of their new well site, to house their workers and their families.

Commemorative marker and wellhead at the exact site of Leduc #1. Leduc #1 Energy Discovery Centre.

Same view to the east along Main Street, toward Telford Lake, Leduc, 1948. (Leduc & District Historical Society)

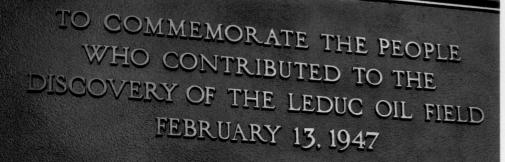

LEDUC #1

"Imperial Oil had drilled 133 consecutive dry wells in search of oil in Alberta and Saskatchewan during a twenty-seven year period and was ready to give up. In 1946, the company decided on one last drilling project – a last chance – at nearby Leduc, south of Edmonton. The wells would be known as "wildcats"– exploratory wells drilled in search of new fields.

The first drill site was Leduc #1. This site was found on a field on Mike Turta's farm, which is located 15 kilometres west of Leduc and about 50 kilometres south of Edmonton. The well was ranked a wildcat. No drilling of any kind had taken place within an eighty kilometre radius. Drilling started on November 20, 1946. At the beginning, the crew thought the well was a gas discovery, but there were signs of something more. Past 1,500 metres the drilling sped up and the first bit samples showed free oil in the reservoir rock.

As a result of this breakthrough, Imperial Oil decided to bring the well in with some fanfare at ten o'clock in the morning on February 13, 1947. The oil company invited civic dignitaries, the media, and the general public to the well site, which is south of what is now Devon. The night before the ceremony, however, swabbing equipment broke down and the crew members laboured all night to repair it. By morning no oil flowed and many of the invited guests left.

Finally, late in the afternoon, the crew prompted the well to flow. Many locals came to see a spectacular column of smoke and fire beside the derrick as the crew flared the first gas and oil. Alberta mines Minister N. E. Tanner turned the valve to start the oil flowing (at an initial rate of about 155 cubic metres per day), and the Canadian oil industry moved into the modern era. Imperial Oil's search for oil finally paid off. By the end of 1947, Imperial Oil and a group of small companies had drilled 147 more wells in the Leduc-Woodbend oilfield. Surprisingly, only eleven were dry.

Commemorative marker at the site of Leduc #1. Leduc #1 Energy Discovery Centre.

Leduc #1 stopped producing in 1974 after the production of 317,000 barrels of oil and 9 million cubic metres of natural gas. On November 1, 1989, Esso Resources – the exploration and production arm of Imperial Oil – began producing the field as a gas reservoir.

When the Leduc #1 blew in. Vern Hunter recalled the event: "By the morning of February 13 – the date set – we hadn't started to swab (a technique for sucking oil to the surface) and that operation sometimes takes days. However, we crossed our fingers and at daylight started in. Shortly before 4 p.m. the well started to show some signs of life. Then with a roar the well came in, flowing into the sump near the rig. We switched it to the flare line, lit the fire and the most beautiful smoke ring you ever saw went floating skywards."   Leduc #1 Energy Discovery Centre

Vern "Dry Hole" Hunter and son Don, 1949.

Some members of the drilling crew, public officials and dignitaries pose in front of 'Leduc #1 in what would become one of the most famous photographs of the Alberta oil industry. February 13, 1947. Photo: Harry Pollard
(Provincial Archives of Alberta, P2733)

This discovery heralded the shift of Alberta's seventy-year-old agricultural economy to a resource-based economy.

All production, which had been in decline, expanded dramatically and the Edmonton area became a petrochemical distribution centre.   www.soulofalberta.com

LEDUC #1 ENERGY DISCOVERY CENTRE

Operated by the Leduc/Devon Oilfield Historical Society, the Leduc #1 Energy Discovery Centre at the Leduc # 1 Historical Site salutes the grit, determination and success of the Alberta oil patch.

Leduc #1 Energy Discovery Centre, Devon, Alberta.

"On February 13, 1947, news of a huge oil strike at Leduc #1 was transmitted around Alberta, Canada and the world. On that cold February day, local residents, government and Imperial Oil officials gathered to watch as Vern "Dry Hole" Hunter and his crew brought Leduc #1 into production. They didn't know it yet, but that well was the start of an oil boom that provided both growth and prosperity to Alberta and all of Canada.

Operated by the Leduc/Devon Oilfield Historical Society, the Leduc #1 Energy Discovery Centre at the Leduc #1 Historical Site salutes the grit, determination and success of the Alberta oil patch and indoor and outdoor exhibits reflect on its history, technology and progress. Here visitors can see where it happened and stand on the floor of a 1940s drilling rig to touch a part of Alberta's rich petroleum heritage. The Centre exists to preserve and display oilfield and related artifacts of the past for the cultural and historical benefit of future generations and to educate individuals about environmental and energy industry related matters."

www.LeducNumber1.com

Leduc #1 Energy Discovery Centre as the rising sun highlights the namesake Leduc #1.

"The Leduc #1 Energy Discovery Centre is the centerpiece of the Leduc #1 National and Alberta Historical Site. The success of Leduc #1 put Alberta and Canada on the world oil map, and forever changed the economy and destiny of Alberta. The Centre is located in Leduc County, 10 minutes west of the Edmonton International Airport.

The Leduc #1 Energy Discovery Centre officially opened February 13, 1997 – 50th Anniversary of the Leduc #1 Discovery Well. The Centre is a 23,000 sq. ft. Museum/Science Centre with Education as its central focus. Eighty acres is open to the public and presents a large number of unique and interactive displays both indoors and outside the Centre. The Centre contains artifacts, photos, scale models and colourful stories – all designed to give the viewing public a taste of what it was like to be a part of our oil history! Working drilling rigs, a service rig, pumpjacks and other current and historical oilfield equipment bring the site to life. Oilfield veterans are frequently on site and are eager to meet with visitors to share the stories of the early years of Canada's energy history.

Purpose
The Centre honours the history, past and present, and represents the future of the Canadian energy industry through displays, models, exhibits, demonstrations, video, publications, actual operations and most importantly, on-site volunteers with a history.

Education for all Ages
The Leduc #1 Energy Discovery Centre is a world class centre for energy education. Knowledgeable staff lead youth and adult groups of all ages on customized discovery tours; available in both of Canada's official languages. Curriculum-matched education programs are available for Kindergarten to Grade 12 and offer up close activities with fossils, rocks, formations, petrochemicals, the flow of matter, history and social studies . . . all the way to economics and trade in today's markets.

Leduc # 1 bathed in a beautiful Alberta sunrise, Leduc #1 Energy Discovery Centre, Devon.

Fuelling Canada's Engine

The energy industry, especially in Alberta, has developed world leaders in exploration, discovery, development, production and technology. Canadian experts from rig hands to geophysicists circle the globe, bringing their technology and cold weather expertise to help develop oil and gas fields world-wide.

The men and women who discovered and developed the petroleum natural resource were and are pioneers. Intuition, invention and innovation characterized these individuals and the industry."

Leduc #1 Energy Discovery Centre – www.leducnumber1.com

Cable tool rig with Leduc # 1 and Energy Discovery Centre in the background, Devon.

View up the inside of Leduc #1's derrick, Devon.

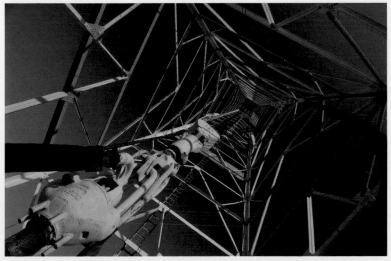

THE TOWN IMPERIAL OIL BUILT

The town of Devon was created in large part by the discovery of oil in Leduc. The people of Leduc were generally fond of having the oil workers in their town, but other citizens were not. Unfortunately, Leduc was a small-scale town and was not as progressive-thinking in that era as it is today.

Immediately after the discovery of oil at Leduc, offices were first established by the Imperial Oil Company in that town. But when the new oil fields rapidly expanded, the company decided to build a new "model town" for its employees and their families. This new town was just north of Leduc #1. Since the oil was discovered in a reef that was formed 350 million years ago in the Devonian Period, the town was called Devon. The chosen site of Devon was a barley field overlooking the North Saskatchewan River; it was considered a good town site.

Row of new homes shortly after construction, ca. 1950.   (Glenbow Archives PA-3689-935
Today many of these homes have survived with matured trees and often renovated for the needs of their current owners.

92

Devon was called Canada's model town – the first community in the country to be planned by a regional planning commission. The Alberta Town Planning Commission was asked to help plan "Canada's First Model Town". From the beginning, all the buildings in Devon had electricity, natural gas, water, and indoor toilets. The houses were pre-fabricated in Calgary, and then put together on basement foundations. The cost of these homes ranged from $3,600 to $6,000. As you can see, housing prices have appreciated in 60 years.

The company launched a bus service to Edmonton and a government ferry was put into operation across the North Saskatchewan River just above Devon. Within three months in 1949, Devon had progressed from a hamlet to a town. Devon had many modern facilities, such as streets, sidewalks, a park, hotels and stores, a curling and hockey rink, a post office, a bank, a golf course and swimming pool, a public library and a theatre. The greatest asset to Devon, however, was Imperial Oil's gas absorption plant, which provided employment for a large number of people.

Townhouse development reflects the appreciation in real estate values since 1949

ATLANTIC #3

Old-timers in the oil patch claim Atlantic #3 was the most spectacular well fire in Canadian history. Flames were 150 metres high and smoke could be seen 150 kilometres away. (Provincial Archives of Alberta. ks 66.1)

"A year after the historic success at Leduc #1, an Atlantic Oil Company rig crew was drilling down to 5,000 feet in the same district. They were reportedly in a hurry and cost-conscious, therefore cut some corners on proven drilling practices. They lost mud circulation and kept drilling without it, thinking they were a safe distance away from the oil and gas zone. Then "all hell broke loose!" On March 8, 1948, Atlantic #3 erupted with a 45-metre-high gusher of gas and oil that splattered the snow.

"The blow out and three-day fire created great interest from the international media."

The shocked men then pumped several tons of drilling mud down the hold and after 38 hours the wild flow was shut off, but not for long. The pressure of the reservoir forced gas and oil through cracks and crevices. Geysers of muck, gas and oil spouted out of the earth in hundreds of craters all over a ten-acre area around the well. More than 15,000 barrels a day of light crude flowed out from the ground for over half a year.

All attempts to control the gusher failed. They repeatedly poured down the hole great quantities of stuff to try to stop the flow: cement, water, sawdust, even feathers.

Six months of a continuous flow of oil and gas up around the derrick resulted in a crater which led to the sinking and collapse of the rig. The toppled rig caught fire on September 6, 1948. It burned for three days as workers bravely fought to put it out. They succeeded, but it was a real learning lesson.

Drilling two relief wells, injecting thousands of gallons of river water and using various plugging devices led to sealing the well permanently on November 9, 1948. About 1.4 million barrels of oil were recovered in a series of gathering pools and ditches. The blowout and three-day fire created tremendous interest from the international media thereby bringing the opportunities of the Alberta oil industry to the attention of the world.

In December, 2006 the location of the Atlantic #3 well was declared a Provincial Historic Site."

soulofalberta.com

North-west corner (today) of 20 to 30-acre area covered by the oil spill and fire following the explosion of Atlantic #3. The mist here is a simple atmospheric ground fog as the rising sun heats the earth.

About two hours east of Calgary, Alberta, the gently rolling prairie grasslands suddenly drop off, plunging the visitor into a world of hoodoos, pinnacles, coulees and buttes. Many who visit these badlands for the first time describe this sudden transition as if they have taken a wrong turn and somehow ended up on the moon. Strange land formations rise up on all sides, sculpted by wind and water into hauntingly beautiful shapes sunbathed in terra cotta, bronze and amber.

A trip to Dinosaur Provincial Park is also a 75-million-year journey back in time. This region was then a subtropical paradise populated by turtles, crocodiles and sharks – and featuring lush vegetation similar to the coastal plains of the south-eastern United States today.

Unique "hoodoo" formations characteristic of the badlands of the Red Deer River Valley east of Drumheller.

HERE, ON THE SHORES OF THE BEARPAW SEA, DINOSAURS ONCE HUNTED AND MATED.

Here, on the shores of the Bearpaw Sea, dinosaurs once hunted and mated – and ultimately met their demise, leaving an amazingly rich fossil and bone record for us to discover today. So far, there have been 35 kinds of dinosaur species unearthed in the park, and it's anyone's guess as to how many more are out there. You can visit a display of excavated bones, covered and preserved exactly where they were found. And throughout the park, you'll also find stakes with numbers on them, each representing a dinosaur find.

The pumpjack at right is slowly filling that blue tank with oil that may have been generated by some of the dinosaurs and sea life from 75 million years ago.

In 1979, Dinosaur Provincial Park was designated a UNESCO World Heritage Site. World Heritage Sites are established for the lasting protection of irreplaceable cultural and natural heritage resources of international significance.

Beautiful Horseshoe Canyon west of Drumheller, Alberta and (above right) pumpjack in the badlands near the hoodoos.

CALGARY

If there's a city in Canada that embodies the "spirit of the west", it's Calgary. The most famous of western rodeos, the Calgary Stampede, is held every year in July and, for 10 days, everybody's a cowboy or cowgirl. (That's a requirement by the way pardner, and there's no getting around it!) Billed as the largest outdoor show on earth, the event celebrates the cowboy way with its centrepiece rodeo featuring bronc and bull riding, barrel racing, roping, and the ever-popular chuckwagon races. The Stampede has long been a tourist attraction and it, along with a thriving ranching industry and the discovery of oil nearby, put Calgary on the map for the rest of Canada. Nestled into the foothills of the Rocky Mountains on its western boundary, the city sits literally at the end of the prairies. Visitors driving into Calgary from the east always marvel at that first view of the skyline as it suddenly reveals itself over a small rise on the Trans-Canada Highway.

The world took notice in 1988, when Calgary staged one of the most successful Winter Olympic Games ever. For the length of those games, a torch burned atop the Calgary Tower and is still illuminated on special occasions. The tower rises 191 metres (621 feet), giving a panoramic view of the entire city; prairie to the east, ranch land to the south and the spectacular Rocky Mountains to the west. Calgary was founded as a North West Mounted Police outpost in 1875. But things changed forever in 1914 when oil was discovered south of the city, and again in 1947 when the giant Leduc field came in near Edmonton to the north. Those discoveries fuelled the growth of Calgary, transforming its agricultural economy almost overnight. Today the city boasts that it is home to the second largest number of corporate head offices in the country and its industrial base, comprised of energy, agriculture, manufacturing, tourism and technology, all contribute to its thriving economy.

Savanna drilling rig #40 working Compton Petroleum well south-east of Calgary.

Thanks in part to escalating oil prices, the economy in Calgary and Alberta was booming until the end of 2008, and the region of nearly 1.1 million people was the fastest growing economy in the country. While the oil and gas industry comprises an important part of the economy, the city has invested a great deal into other areas such as tourism and high-tech manufacturing. Over 3.1 million people now visit the city annually for its many festivals and attractions, especially the Calgary Stampede. The city has ranked high in quality-of-life surveys: 25th in 2006, 24th in 2007, 25th in the 2008 and 26th in 2009 Mercer Quality of Living Survey, and 10th best city to live in according to the Economist Intelligence Unit (EIU). Calgary was ranked as the world's cleanest city by Forbes Magazine in 2007. And as the first decade of the 21st century drew to a close the price of a barrel of black gold recovered and the stampede city was back in the game.

Calgary skyline with its signature Saddledome (left foreground) and the Calgary Tower (centre).

Calf ropers in action, Calgary Stampede.

CALGARY AIRPORT (YYC)

WestJet 737 on short final for runway 16.

Calgary's airport had seen improvements in lighting, facilities and an expansion of the offices of Trans Canada Airlines (today's Air Canada) after the war. In 1949, the City of Calgary took back control of the airport from the federal Department of Transport.

Small regional carriers in areas like Grande Prairie and Peace River tried to gain a foothold, but found it very difficult and did not continue. There were a number of reasons for this, one being that private oil and mining companies operating in the north bought their own aircraft and hired their own pilots.

Renowned Canadian pilot W. R. "Wop" May standing in front of the automobile with an Imperial Oil aircraft in background, Edmonton Airport, December 1920. Like many other Canadian WWI Royal Air Force (RAF) pilots who returned after the war, Wop May established a civilian aviation company. May Airplanes limited was formed in 1919 with the cooperation of the City of Edmonton.
(McDermid Photo, Glenbow Archives NC-6-6055)

DC-3 of Trans Canada Airlines (today's Air Canada) being fueled at Calgary Airport. ca. 1950 (Provincial Archives of Alberta P 2779)

The Calgary International Airport has come a long way since its inception in 1914. Situated in Bowness, 10 kilometres from the city, the original airfield was comprised of a grass airstrip and a ramshackle hut, which served as both hangar and terminal building.

In 1938, the facility moved to its present location in Calgary's north-east and was christened McCall Field, in honour of Captain Fred McCall, a World War I flying ace and one of Calgary's pioneer aviators.

Over $1 billion has been invested since 1992 to renovate and expand airport infrastructure and a further $3 billion investment is currently forecast over the next 10 years to meet the needs of the trading region. A new International Terminal Building, new parking structures and a 14,000 foot parallel runway are just some of the initiatives in place that will ensure that the Calgary International Airport remains a premiere global gateway capable of accommodating a significant increase in passengers and also, the new and larger aircraft of the future.

Airside, Calgary Airport today.

ALBERTANS LOST IN OILPATCH TRANSLATION

By Deborah Yedlin, Calgary Herald – March 23, 2010

Many people probably don't realize that one in six jobs in Alberta is tied to the energy sector. With that kind of weighting – not unlike Ontario's dependence on the auto sector – it's not hard to see what can happen to Alberta's economy if the energy sector falls on hard times.

Many may not be aware that the energy sector makes up about half of the provincial economy, with every dollar spent generating $3 for the broader economy.

Here's what's really troubling: if Albertans don't "get it," then it stands to reason even less is understood outside the province.

When one looks at the impact of the energy sector on a national basis, the economic impact is a whopping $110 billion, and yet Ontario and Quebec are quick to disparage the industry even though they benefit in transfer payments.

Some company executives, such as Canadian Oil Sands Trust chief executive Marcel Coutu, are so frustrated with the communication gap they have taken it on themselves.

Since last fall, Coutu has made numerous presentations – primarily in Eastern Canada – aimed at de-constructing the perceptions held by Canadians about the oilsands, emphasizing the strides that have been made, the improvements in technology and the economic benefit that extends far beyond Alberta's borders.

Wherever possible, Coutu has spoken at chambers of commerce, held town hall meetings and met with university students.

And while he admits he doesn't always get the turnout he hopes for, those who do show up leave with a much better understanding of what is a major economic driver in this country.

Coutu's "man to man" coverage, however, is a far cry from what happens south of the border, where every Sunday morning the political talk shows are filled with messages from the energy sector about the difference it makes to the economy – as well as the strides being made from an environmental perspective.

The cynic might say it's all self-serving.

Not entirely.

Just like the disappearance of the energy sector would mean a sales tax of 16 per cent in Alberta to cover for the lost revenue, the same would be true south of the border.

And let's not even start with what it would mean to federal transfer payments.

Alberta is vulnerable enough to the dynamics of the global economy – the variables that can be controlled are the ones that fall under the purview of the provincial government, namely the underlying fiscal structure that governs investment decisions by the private sector.

Since Alberta does not have a consumption tax, it's even more important the energy sector is vibrant; otherwise there's no money for what Albertans expect from their government.

Success for the energy sector is good not just for Albertans, but also for the health of the Canadian economy. It's time that message was more effectively delivered – and by more than one individual.   www.calgaryherald.com

This article first appeared in the Calgary Herald and was used with the permission of Canwest Publishing Inc.

SUCCESS FOR THE ENERGY SECTOR IS GOOD NOT JUST FOR ALBERTANS, BUT ALSO FOR THE HEALTH OF THE CANADIAN ECONOMY.

Suncor refinery, Edmonton.

CLIMATE CHANGE

Climate change is a wide-scale change in average weather over a time period of at least 30 years. Climate change can be caused by a number of factors, such as changes in the Earth's orbit, volcanic eruptions, or changes in energy from the sun.

Greenhouse gases have an important effect on Earth's temperature. They trap heat in the atmosphere and cause global temperatures to rise. This is called "the greenhouse effect". Naturally occurring greenhouse gases (GHGs) are essential for our survival; they act like a blanket around Earth, trapping heat in the lower layer of our atmosphere and prevent most heat from escaping.

The issue is that humans have substantially increased the amount of naturally occurring GHGs by burning fossil fuels, including coal, oil and natural gas. Scientists now agree that human activity is most likely responsible for most temperature increases over the past 250 years. The biggest concern is the speed at which these changes are happening.

Carbon dioxide is the main concern. Information shows that atmospheric levels of carbon dioxide are increasing by more than 10 per cent every 20 years. If emissions continue to grow at current rates, the level of atmospheric carbon dioxide will almost double during the 21st century; it's possible it could even triple.

However, we can mitigate and reduce the effects of climate change. By protecting, conserving and enhancing our wetlands, forests, and other natural spaces, we help preserve carbon sinks. We can also reduce our energy consumption and commit to actions that help conserve our natural resources.

www.environment.alberta.ca

Fall colour in Kananaskis Country.

Kananaskis golf course.

CARBON FOOTPRINT

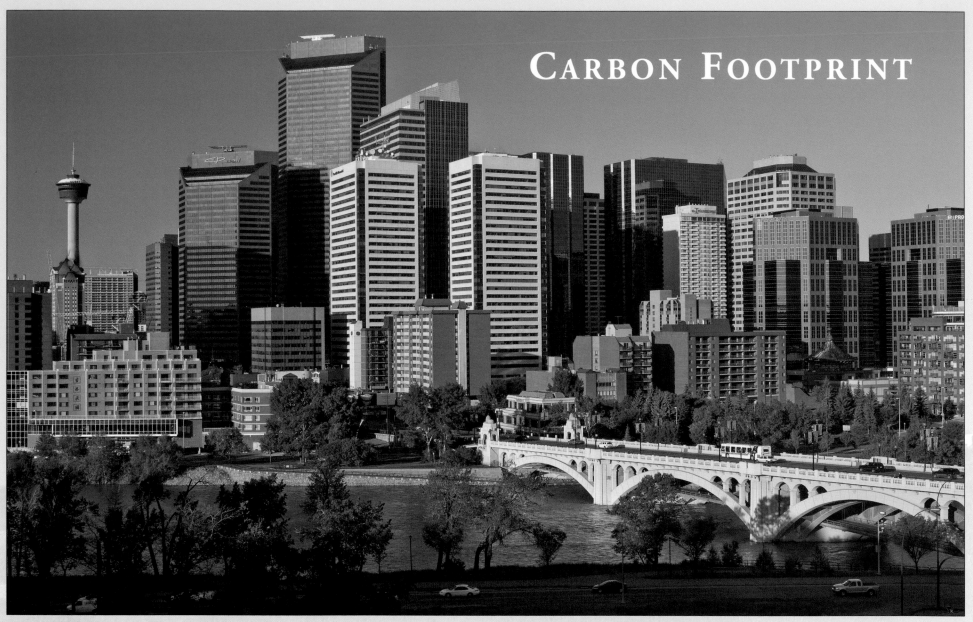

View of a thriving, downtown Calgary and the Bow River.

What Is It?

It's a term that's being thrown around media circles like a volleyball on the beach: carbon footprint. It's a measure of the affect you have on the environment.

Specifically, your carbon footprint is a barometer of the amount of greenhouse gases your produce in your daily activities. There are two levels of carbon footprints: primary and secondary, and they're both measured in units of carbon dioxide (or CO_2).

According to an article titled "Carbon Footprint of Best Conserving Americans is Still Double Global Average" in Science Daily (sciencedaily.com, accessed March 2010), the average carbon footprint in North America is about 20 metric tons per year. The world average (including the United States) is 4 tons, according to the article.

Primary and Secondary Footprints

Your primary footprint is the amount of greenhouse gases you use from burning fossil fuels for energy and transportation. Your secondary footprint is for indirect greenhouse gas emissions from using products (such as plastic water bottles).

Why is This Important?

Most scientific evidence indicates that greenhouse gas emissions contribute to global warming. This can be seen most drastically by looking at climate change. According to Carbon Footprint, a carbon footprint consulting service, the majority of greenhouse gas emissions comes from humans in the forms of carbon dioxide, methane and nitrous oxide. Changes in the earth's various climates could be detrimental to all life on earth. If the polar ice caps melt, for example, the ocean-water levels will rise and liveable land mass on all of the seven continents will decrease. With unchecked climate change, the temperature on earth could, potentially, reach a level that is uninhabitable all together.

Carbon Footprint allows you to calculate your carbon footprint on a variety of levels (from how much electricity you use to how often you drive your car to the kinds of foods you eat). It is best to use on-line tools to accurately assess the amount of greenhouse gases you are directly and indirectly producing.

On the Road:
- Reduce the amount you drive.
- Walk, bike, carpool and use public transit more often.
- Don't idle your vehicle.
- Consider fuel efficiency when buying a vehicle.

At Home:
- Buy energy-efficient appliances.
- Shut off lights and equipment when not in use.
- Use fluorescent bulbs, lower your thermostat and water heater.
- temperature.
- Insulate and weatherproof your home.
- Conserve water.
- Reduce, reuse, recycle.

At Work:
- Think twice about printing
- If you cannot reuse paper, use the recycled kind.
- Buy energy-saving office appliances and equipment, such as EnergyStar-approved computers, LCD monitors, printers and photocopiers.
- Set up a recycling program for used paper, aluminum and other materials.
- Fix leaking taps right away.
- Install devices in taps and showers that reduce the water volume.
- Turn lights on only when needed.
- Set the thermostat at 19 degrees Celsius.
- If you can't measure it you can't manage it. Check your use of electricity, gas and oil, so you know how much energy your office uses and how much you can reduce.

www.lisanyren.com

Drive, ride public transit, fly, rollerblade, pedal or walk.

Carbon Capture & Storage (CCS)

Kananaskis River. Founded in 1977, Kananaskis Country is a network of parks and protected areas, recreation areas and multi-use areas with cattle grazing, oil and gas extraction and logging.

What is CCS?

Carbon capture and storage (CCS) is a process that captures carbon dioxide (CO_2) emissions and stores them in geological formations deep inside the earth.

CO_2 contributes to greenhouse gas emissions (GHGs), the bulk of which come from the production and use of fossil fuels – coal, oil and gas – as well as electricity. CCS also has the potential to reduce emissions from Alberta's value-added and manufacturing industries, such as petrochemical development.

What's happening with CCS in Alberta?

On April 24, 2008, Premier Ed Stelmach announced the creation of the Alberta Carbon Capture and Storage Development Council. The council brings together experts from the public and private sectors. Setting up the council was a commitment made in Alberta's 2008 Climate Change Strategy. Under the strategy, Alberta committed to reducing projected emissions by 200 megatonnes by 2050 – 70 per cent of which will be achieved through CCS.

On July 8, 2008, the Alberta Government announced it will contribute $2 billion to reduce GHG emissions through new CCS projects. The expected result is five million tonnes in annual reductions by 2015— comparable to taking one million vehicles off the road.

Is CCS safe?

Experience in Canada and around the world has shown that CCS can be done safely and produce positive environmental results. In Alberta, porous sedimentary rock formations beneath non-porous formations are ideally suited for the injection and long-term, safe and secure storage of carbon dioxide. The CO_2 will be separated from other emissions, then dehydrated, compressed and transported by pipeline to a storage site where it will be injected one to two kilometres deep into the porous rock formation. It will then be sealed and monitored by experts to ensure there is no leakage or impact on either public safety or the environment.

Does CCS work?

EnCana's CO_2 project in Weyburn is the world's largest, full-scale scientific field study of its kind. This encouraging CCS pilot project pipes CO_2 from Beulah, North Dakota to Weyburn, Saskatchewan where it is injected into a depleted oil field. Since 2000, more than 13 million tonnes of CO_2 have been injected into the oil field with no adverse effects. The goal is to store 30 million tonnes of CO_2. This equates to the emissions of 6.7 million cars over one year. An international team of scientists has detected no leakage of CO_2 after extensive monitoring of the project.

Why is the Alberta government investing in CCS?

CCS provides an opportunity for the province to reduce GHGs while ensuring Alberta's and Canada's economic success and growth can continue.

This provincial investment is intended to accelerate the development of projects and encourage the necessary investment from industry to make CCS viable in Alberta. Industry produces commodities that Albertans use, including electricity, natural gas and oil. The products, such as oil, that are exported to other jurisdictions represent the core of Alberta's economy, and create jobs, royalties and taxes.

An investment in CCS is also an investment in the environment. CCS in Alberta has the potential to be North America's largest single source of GHG emissions reductions.

CCS demonstrates Alberta's action to reduce emissions while continuing to be a major supplier of resources that are developed in an environmentally responsible manner.

Government of Alberta
www.energy.gov.ab.ca

VAST DEPOSITS OF OIL, GAS AND COAL THAT TOOK HUNDREDS OF MILLIONS OF YEARS TO ACCUMULATE ARE BEING EXTRACTED AND BURNED IN THE GEOLOGICAL BLINK OF AN EYE.

The following report from the Science and Technology Division of the Parliamentary Information and Research Service of the Library of Parliament (Canada) was published in 2006.

Vast deposits of oil, gas and coal that took hundreds of millions of years to accumulate are being extracted and burned in the geological blink of an eye. The rapid release of carbon dioxide (CO_2) from the combustion of fossil fuels has led to a build-up of this gas in the atmosphere. This in turn is affecting how plants grow and use water, causing the oceans to acidify and, most importantly, causing an imbalance in the manner by which solar energy is trapped in the atmosphere.

Anthropogenic climate change is now seen by most governments as something that needs to be addressed. Technologies are available that can reduce emissions, but none has the capacity to address the problem single-handedly. One of the options that is gaining interest is carbon capture and storage.

The only real long-term way of reducing greenhouse gas emissions, assuming that energy use will continue to grow, is to uncouple energy use and CO_2 release. This can be done by moving to low-carbon-energy sources such as hydro, nuclear, wind, solar, geothermal and tidal energy. However, currently available low-carbon-energy sources and technologies cannot fully substitute for fossil fuels, particularly given the numerous ways in which such fuels are used and the current economic framework for energy.

The result is that fossil-fuel-dependent energy infrastructure is currently being planned that will remain for decades, continuing to increase atmospheric CO_2. It has been estimated that lifetime emissions from power plants projected to be built for the next 25 years are equivalent to all emissions for the last 250 years.

An alternative possibility, however, is that fossil fuels can continue to form the basis of our energy infrastructure, but that CO_2 can be captured from the emissions and stored away from the atmosphere for very long periods of time. The atmosphere has been a convenient repository for such wastes, but there is no technical reason why this has to be. Other parts of the Earth, notably geological formations and the oceans, can also be used to store CO_2.

Before carbon can be stored in a manner that prevents it from entering the atmosphere, it must first be removed from the emissions emanating from the burning of fuel. Removal of CO_2 from widely distributed and/or mobile emission sources such as residential heating, automobiles and airplanes, is impractical but must be addressed.

If fuels for mobile and distributed energy needs continue to be carbon-based, then CO_2 must be dealt with after the emission has taken place. This could be through increasing biological sinks (e.g., forests), or indirectly through displacing fossil-fuel-based electricity with carbon-free sources such as wind.

The latter option is likely to be the most viable, since biological sinks are limited in capacity. Other technological approaches such as capturing atmospheric CO_2 in carbon-absorbing chemicals have been suggested but not as yet demonstrated.

Carbon capture and storage is mostly used to describe methods for removing CO_2 emissions from large stationary sources, such as electricity generation and some industrial processes, and storing it away from the atmosphere.

Capture

There are three basic methods for capturing CO_2 from such emission streams: pre-combustion; post-combustion; and oxyfuel combustion.

Pre-combustion reacts a primary fuel with air or oxygen and steam to produce hydrogen and carbon monoxide which can be further treated to produce more hydrogen and CO_2 (15 per cent-60 per cent by volume), which can then be separated. The hydrogen (synthesis fuel) can be used for energy, with water as a by-product; the CO_2, at relatively high pressure and concentrations, is more amenable to capture than would have been the case if the fuel were combusted as is. This type of technology is used in Canada in four coal-fired integrated gasification combined cycle (IGCC) plants, although none capture the CO_2. The production of synthesis gas is relatively old technology but has been used in combination with pre-combustion capture only in specific circumstances.

Post-combustion processes generally use a recyclable solvent to trap CO_2 in the emission stream, though some projects are attempting to demonstrate biogenic capture through photosynthetic algae. Post-combustion capture has been used only in specific circumstances.

Oxyfuel combustion burns the primary fuel in almost pure oxygen to produce a very pure CO_2 stream which then can be compressed for storage purposes. This process, however, requires a fairly elaborate mechanism to purify the oxygen. Oxyfuel technology is in the demonstration phase. In other cases, such as fertilizer production (ammonia), CO_2 is separated as part of the chemical process of producing the product. These latter types of operations operate already in a mature market, though once again not in combination with storage.

Morning rush hour southbound on the Deerfoot Trail (Highway 2) in Calgary.

All technologies as applied to energy generation effectively reduce efficiency, increasing the amount of CO_2 created (and therefore necessary to capture) per unit of energy produced.

Capturing CO_2 is the most energy-intensive phase (and therefore the largest contributor to CO_2 releases and costs of energy production) in a complete carbon capture and storage mechanism.

It should also be noted that since biofuels such as ethanol and biodiesel are derived from plant material that uses atmospheric CO_2, the capture and storage of CO_2 from the combustion of these fuels would actually remove CO_2 from the atmosphere.

Transportation

Many point sources of captured CO_2 would not be close to geological or oceanic storage facilities. In these cases, transportation would be required. The main form of transportation envisioned is pipeline, though shipping would be a possibility should there be a need. Trains and trucks are thought to be too small-scale for projects of this size.

Pipelines would require a new regulatory regime to ensure that proper materials are used (CO_2 combined with water, for instance, is highly corrosive to some pipeline materials) and that monitoring for leaks and health and safety measures are adequate. However, these are all technically possible, and pipelines in general currently operate in a mature market.

Storage

Storage of CO_2 is mostly discussed in terms of geological storage, though oceanic storage has also been noted as a possibility. Geological storage can take place in oil and gas reserves, deep saline aquifers and unminable coal beds. The injection of CO_2 at pressure into these formations, generally at depths greater than 800m, means that the CO_2 remains a liquid and displaces liquids, such as oil or water, that are present in the pores of the rock.

Liquid CO_2 is lighter than water and therefore tends to travel upward; thus, suitable geological formations must have a "cap" rock to act as a barrier to its movement. If the cap rock is insufficiently wide, CO_2 could leak out around the edges. In this case, mechanisms would be required to prevent such leakage. The viability of any such project would have to be established on an individual basis.

There is also the possibility of injecting CO_2 into the deep ocean. CO_2, however, reacts with water to produce carbonic acid, which makes the water more acidic. Many aquatic organisms are highly sensitive to changes in acidity, making oceanic storage more problematic than geological storage from an ecological standpoint.

Estimated Capacity

One of the main goals of the IPCC study was to examine the global capacity for carbon storage. The IPCC report concluded that, from a technical standpoint, there is a lower limit of about 1,700 Gigatonnes (Gt; one Gt equals one thousand million tonnes) of CO_2 capacity, with very uncertain upper limits (possibly over 10,000 $GtCO_2$). From a more realistic (economic) point of view, however, the report suggested that there is almost certainly a minimum of 200 $GtCO_2$ capacity and likely 2,000 $GtCO_2$.

At the lower estimate, all current emissions could be sequestered for approximately nine years, and for 90 years at the upper estimate. Of course, by those dates the definition of what is economically viable and/or the need for capture could be very different.

There are currently three major carbon storage operations in the world, storing approximately four megatonnes of CO_2 per year (0.004 Gt/year). One of these is the Weyburn Canada enhanced oil recovery project in Saskatchewan. The project increases oil recovery by pumping CO_2 (from the United States) into the oil field. The CO_2 remains in the ground, thereby displacing oil and facilitating its extraction.

Conclusion

Technologies are available that can make incremental and very important contributions toward reducing greenhouse gas emissions over the next few decades. The IPCC concluded in its special report on carbon capture and storage that this technology is one of them. The capacity is significant, if not well constrained. Geological storage is preferred over oceanic storage, as it is a more mature technology with fewer ecological implications. Each project, however, must be examined for its viability, in particular with regard to limiting leakage rates to acceptable levels. In an extreme case, the burning of biofuels and CO_2 would lead to removal of CO_2 from the atmosphere.

Capture technology alone increases energy inputs from 10 per cent to 40 per cent, which would be reflected in energy costs to the consumer. Any significant implementation of carbon capture and storage, therefore, will require public support in order to create the political will to act.

Science and Technology Division of the Parliamentary Information and Research Service of the Library of Parliament (Canada) was published in 2006.

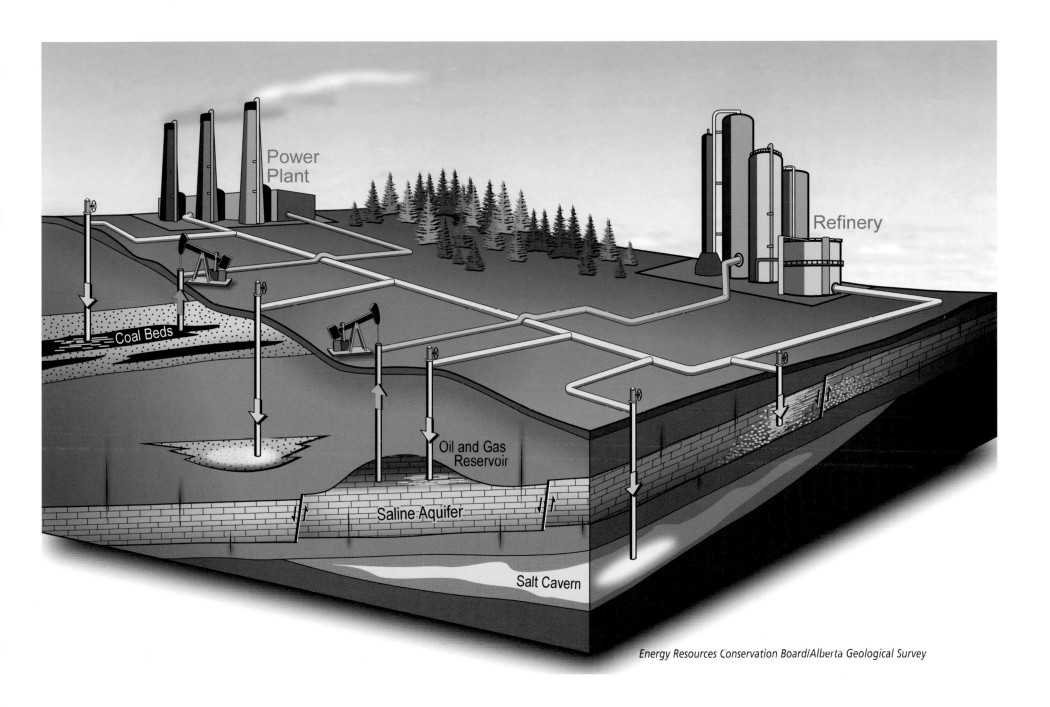

Power Plant

Refinery

Coal Beds

Oil and Gas Reservoir

Saline Aquifer

Salt Cavern

Energy Resources Conservation Board/Alberta Geological Survey

CARBON CAPTURE'S FUTURE

In July of 2008, the Government of Alberta announced it would invest two billion dollars in Carbon Capture and Storage (CCS) to help reduce greenhouse gas (GHG) emissions. The Government requested Expressions of Interest and the projects that most closely fit the guidelines were asked to submit Full Project Proposals. The goal of the funding is to reduce emissions at facilities such as coal-fired electricity plants and oil sands upgraders by up to five million tonnes annually, starting in 2015.

The following four projects have been approved and are proceeding.

1. **Swan Hills Synfuels** is adapting existing *in-situ* coal gasification technology to turn deep, stranded coal into clean synthesis gas (syngas) in Alberta.

2. **Enhance Energy and North West Upgrading's** Alberta Carbon Trunk Line (ACTL) is a 240-kilometre(150-mile) pipeline that, when completed will employ safe proven technology to gather, compress and store up to 14.6 million tonnes of CO_2 per year at full capacity.

3. **TransAlta Project Pioneer** will utilize leading-edge technology to capture CO_2 which will be used for enhanced oil recovery (EOR) in nearby conventional oil fields, or stored almost three kilometres underground.

4. **Shell Quest Project** will capture and store 1.2 million tonnes of carbon dioxide annually beginning in 2015 from Shell's Scotford upgrader and expansion, near Fort Saskatchewan.

Swan Hills Synfuels is adapting existing *in-situ* coal gasification technology to turn deep, stranded coal into clean synthesis gas (syngas) in Alberta. This clean syngas is an environmentally responsible energy source that can be used as fuel for clean power generation or processed further to create other clean energy products.

Swan Hills Synfuels, a private Alberta developer, is focused on the development of clean energy projects utilizing *in-situ* coal gasification (ISCG) technology combined with carbon capture and storage. Swan Hills Synfuels has secured reserves of deep coal and is tapping this coal's energy with ISCG to develop clean energy projects in Alberta.

ISCG, combined with carbon capture and storage, will create a new clean energy industry in Alberta. Environmentally, ISCG with carbon capture and storage has an exemplary profile including significantly reduced air emission, non-fresh water use in the gasification process and minimal surface disturbance.

Swan Hills Synfuels' first major commercial project is the Swan Hills *in-situ* Coal Gasification and Sagitawah Power Generation Project (Swan Hills ISCG/Sagitawah Power Project) located in Central Alberta. This clean energy system will produce syngas that will be used as fuel for very efficient low-emissions power generation, producing 300 MW of baseload electricity supply for Albertans. The project will also capture and sequester over 1.3 million tonnes of CO_2 each year. This project is scheduled to be operational in 2015.

Albertans will benefit from this project on a number of levels: clean energy, secure source of electricity, reduced CO_2 emissions, job creation and optimally located facilities. In addition, with the planned use of the project's captured CO_2 for enhanced oil recovery (EOR), Albertans will receive incremental benefits from this EOR activity, such as increased oil royalties.

The first round of commercial scale projects is expected to achieve annual carbon dioxide reductions by 2015 equivalent to taking approximately one million vehicles, or about a third of all registered vehicles in the province, off of the road.   energy.gov.ab.ca

Enhance Energy is building the Alberta Carbon Trunk Line (ACTL), a 240-kilometre pipeline that will transport CO_2. In 2010 this was the only project with clear plans to use stored CO_2 for enhanced oil recovery (EOR). The initial supplies of CO_2 will come from the Agrium Redwater Complex and once built, the North West Upgrader. North West Upgrading will upgrade bitumen from Alberta's oil sands and the captured CO_2 will be transported to depleting conventional oilfields and used in enhanced oil recovery.

Specializing in enhanced oil recovery and the storage of carbon emissions, Enhance Energy is a Calgary based company currently building the world's largest carbon capture and storage project. The ACTL pipeline employs safe proven technology to gather, compress and store up to 14.6 million tonnes of CO_2 per year at full capacity. This will be equivalent to removing 2.6 million cars off the road annually or about a third of all registered vehicles in the province. To add to this advantage, the stored CO_2 will be injected into depleted oil reservoirs and result in the recovery of over 1 billion barrels of oil, thereby increasing EOR recovery of original oil in place by 10 to 15 percent. This will translate into more than $15 billion in royalties for Alberta.

The CO_2 pipeline will reduce carbon footprints for upgraders and refiners that feed into the system and holds much promise for the future of the oil and gas industry in Alberta.

At the Keephills 3 plant west of Edmonton, TransAlta Corporation and its partners in Project Pioneer will utilize leading-edge technology to capture CO_2 which will be used for enhanced oil recovery (EOR) in nearby conventional oil fields, or stored almost three kilometres underground. The project is expected to capture one million tonnes of carbon dioxide annually beginning in 2015.

The Alberta government and TransAlta Corporation are working together to kick-start a leading-edge coal-fired electricity generation by perfecting technology to reduce greenhouse gas emissions from coal-fired electricity plants. "This will not only have a significant impact in Alberta, but it could help throughout North America and in developing nations like China," said Premier Ed Stelmach. "This project provides an opportunity for Alberta to be a leader in developing game-changing carbon capture technology that could be used around the globe."

Project Pioneer represents a major step toward advancing the capture of greenhouse gas emissions. The project entails the construction of a large carbon dioxide (CO_2) capture and storage (CCS) facility at one of TransAlta's coal-fired generating stations west of Edmonton, Alberta. Once complete, Project Pioneer will be one of the largest CCS facilities in the world and the first to have an integrated underground storage system.

A report by the United Nations Intergovermental Panel on Climate Change (IPCC) states CCS could contribute up to 55 per cent of the emission reductions scientists believe are necessary to address global warming. CCS would allow modern societies to manage the CO_2 emissions from conventional fossil fuels even as they develop and deploy renewable energy sources and energy efficiency measures. CCS by itself will not solve the climate challenge, but it could be an important part of Alberta's strategy for reducing CO_2 emissions from oil sands.

Shell upgrader north of Fort McMurray with Muskeg River mine in the background.

Shell, on behalf of the Athabasca Oil Sands Project, a joint venture among Shell Canada (60 per cent), Chevron Canada Limited (20 per cent) and Marathon Oil Sands L.P. (20 per cent) has proposed a carbon capture and storage (CCS) project. The Quest CCS Project would be based at Shell's Scotford Upgrader, located near Fort Saskatchewan, Alberta. Commissioned in 2003, the Upgrader turns bitumen from the Athabasca oil sands into synthetic crude oil, most of which is turned into consumer products such as gasoline.

The Scotford Upgrader is located next to Shell Canada's Scotford Refinery near Fort Saskatchewan, Alberta. The Scotford Upgrader uses hydrogen-addition technology to upgrade the high viscosity "extra heavy" crude oil (called bitumen) from the Muskeg River Mine (photo at left) into a wide range of synthetic crude oils. A significant portion of the output of the Scotford Upgrader is sold to the Scotford Refinery. Both light and heavy crudes are also sold to Shell's Sarnia Refinery in Ontario. The balance of the synthetic crude is sold to the general marketplace.

This upgrader, which is part of the joint venture project between Shell Canada, Chevron Canada (a wholly owned subsidiary of Chevron Corporation) and Marathon Oil Sands L.P., is operated by Shell Canada.

Upgrading is the process of breaking large hydrocarbon molecules (such as bitumen) into smaller ones by increasing the hydrogen to carbon ratio. These upgraded crude oils are suitable feedstocks for refineries which will process them into refined products like gasoline.

Scotford's upgrading process adds hydrogen to the bitumen, breaking up the large hydrocarbon molecules, this process is called hydrogen-addition or hydrogen-conversion.

www.shellcanada.ca

Shell Canada's Scotford Upgrader and Refinery near Fort Saskatchewan.

Drillers Martin Hovis and Joseph Brown at Dingman Discovery Well, 1914.
(Provincial Archives of Alberta, P1303)

Ottawa's National Energy Program (NEP) scuttles a decade of progress in Canada's petroleum industry

The road to hell is paved with good intentions. And the National Energy Program (NEP) had plenty of good intentions: protect Canadian consumers from escalating energy costs, promote the country's oil self-sufficiency and maintain the oil supply – particularly for the industrial base in eastern Canada. Promote Canadian ownership of the energy industry, exploration for oil in Canada, and alternative energy sources. Increase government revenues from oil sales through a variety of taxes and agreements.

The October 1980 federal budget was really less a federal budget than purely an energy policy statement. The budget held virtually nothing to address inflation and recessionary trends in the economy, while the NEP changed the structure of the Canadian petroleum industry and put an end to a decade of industry momentum towards self-sufficiency.

At the heart of the energy program was a system of price controls keeping domestic oil prices below those of the world market, export taxes on petroleum products, and an alphabet soup of incentives ranging from tax concessions to outright grants aimed at accelerating the search for oil and gas on federal lands, largely in the north and offshore the east coast.

The industry reacted by cutting activity levels across Canada. In Alberta, the number of wells drilled plunged from more than 7,000 in 1980 to just over 5,000 in 1982, sparking an exodus that saw the rig fleet in the province plunge to 154 in 1983 from nearly 340 in 1980.

To show his displeasure, Alberta premier Peter Lougheed asked the legislature to approve a 15 per cent cut in crude oil production as a "measured response" to the "discriminatory and unfair" federal budget. The first round of cuts – about 60,000 barrels per day – was implemented in March 1981.

It is estimated that Alberta lost between $50 billion and $100 billion because of the NEP, while through the dark years of the program, 1980 to 1985, the bankruptcy rate in Alberta rose by 150 per cent – three times the national average.

Oilweek, June 2008 – 60th Anniversary Edition

Doom and gloom
There were precious few bright spots in the oil patch in 1981

Budgets throughout the industry were slashed following the federal government's introduction of the National Energy Program in October of 1980.

In January 1981, Syncrude Canada announced it would suspend a planned $2 million expansion in the oilsands. Mobil Oil Canada cut its budget in the same month by a whopping 46 per cent and decided to spend only enough to meet appraisal and exploratory commitments in the Sable Island area and on the Grand Banks.

The exodus of rigs from western Canada continued in February, with the count reaching 150 by the middle of the year as Canadian oilmen realized the advantages of operating in the United States.

In March of 1981, the Alberta government brought in the first of its three stages of cutbacks in light and medium crude production that would eventually see output cut by 180,000 barrels per day.

The provincial government's take from six land sales in the first quarter of 1981 dropped by 70 per cent from the first quarter of 1980. But amidst the general doom and gloom, there were bright spots. Most notably these came in frontier exploration.

Mobil Oil Canada got the ball rolling with continued encouraging results from Hibernia B-80, a step-out to its Hibernia discovery well. The well tested oil from all pay zones encountered and cumulative production testing totalled nearly 19,000 barrels per day.

And with both the High Arctic and the Beaufort Sea recording three strike, north of 60 had a tremendous year, largely supported by Ottawa's extensive suite of NEP incentives.
Oilweek, June 2008 – 60th Anniversary Edition
www.oilweek.com

Tyler Rook, Brandon Erdos and Mark Shore working the drilling floor of Nabors rig #19 near High River.

ROYALTIES

Facts on Royalties (Government)

Royalties are an important part of the Alberta government's revenue stream. They help fund important programs like health, education and infrastructure. They ensure that Albertans receive a portion of the benefits arising from the development of the province's energy resources. A well-designed royalty system endeavours to strike the right balance between returning a share of the profits to the province as resource owner, while encouraging risk taking by the private sector to develop the resource which creates jobs and economic growth.

The Alberta Department of Energy is responsible for the administration of the Mines and Minerals Act which sets out the requirements for the responsible development of Alberta's non-renewable mineral resources. In Alberta, 81 per cent of the subsurface mineral rights are owned by the Crown. The remaining 19 per cent are owned by the Government of Canada in national parks or held on behalf of First Nations and by individuals or corporations as a result of land grants made by Canada in the 1800s.

Companies are granted the right to explore for and develop petroleum and natural gas resources in exchange for the value to Albertans that flows from development in the form of royalties, bonus bid payments (the amount of money offered or bid for the mineral rights) and rents. A royalty is the price charged by the energy resource owner for the right to develop those resources.

Between 2006 and 2009 royalties averaged about $10 billion for each of those three years. With the dramatic drop in the price of a barrel of oil in late 2008, these figures are expected to drop by about 50 per cent.

Royalties for conventional oil are set by a sliding rate formula containing separate elements that account for oil price and well production. Royalty rates will range up to 40 per cent, with rate caps at $120 per barrel (bbl).

Gas royalties are set by a sliding rate formula sensitive to price and production volume. New royalty rates will range from 5 per cent to 50 per cent with rate caps at $17.75 Cdn/GJ (gigajoule).

To advance Alberta's competitiveness in the upstream oil and gas sector, the Alberta government will modify conventional oil and natural gas royalty rates; promote more innovation and use of new technologies and reduce unnecessary red tape while improving coordination of regulatory processes.

"Alberta is recognized around the world as a leader in technical innovation and development of energy resources," said Premier Ed Stelmach. "These changes support the Provincial Energy Strategy. They will help us use innovation to unlock our energy resources, create opportunities and jobs in communities large and small across our province and strengthen Alberta's economic recovery."

In recent years, new drilling technologies have unlocked sizable new reserves in other North American jurisdictions that are now competing for investment. "We can't pretend that oil and gas investment levels haven't eroded or that we don't have a responsibility to current and future generations of Albertans to address that," said Energy Minister Ron Liepert. "Being competitive has a positive impact far beyond the energy sector. It contributes greatly to our communities, our standard of living, and our prosperity as a province and as individuals."

Currently almost one in seven Albertans are directly or indirectly employed by the energy industry. Changes to improve Alberta's competitiveness are expected to create 8,000 jobs in 2011-12 and then 13,000 more jobs annually across the economy. Over the next 25 years conventional oil and gas development in Alberta has the potential to add $1 trillion in new economic activity while the oil sands could add another $1.5 trillion.

"We need to do more to explain to Albertans the ways in which our energy sector drives our economy. Albertans need to have factual and balanced information about how energy development happens in Alberta and just how critical it is to meeting our economic goals," said Liepert.

Adapted from: www.energy.alberta.ca

Facts on Royalties (Industry)

After three years of acrimony between the energy industry and the provincial government, Alberta finally delivered its long-awaited changes to the royalty structure in May of 2010.

The Canadian Association of Petroleum Producers' (CAPP) reacted favourably with a press release stating that; "today's Alberta government royalty announcement largely delivers on the positive direction established with the release of the competitiveness review report in March.

"The new fiscal details are particularly positive for the competitiveness of Alberta's natural gas and will enhance the industry's ability to strengthen the economy and create jobs for Albertans," said CAPP president David Collyer. "On the oil side, changes are not as significant. We are encouraged by the broader application of the lower up-front royalty rate, which will stimulate new oil drilling activity."

A review of Alberta's position as a competitive place for oil and gas investment relative to other North American jurisdictions was completed in March 2010. The review considered the role of Alberta's fiscal regime, the regulatory framework, technology and innovation and the overall business climate. As a first step to improving Alberta's oil and gas investment climate, maximum royalty rates were lowered and an existing up-front five per cent royalty rate feature was made permanent.

CAPP believes the Government of Alberta's announcement of fiscal details in May of 2010 is helping to restore investor confidence in Alberta's oil and gas industry. Increased oil and gas investment in Alberta translates to increased activity, jobs and public revenues, all of which contribute to the quality of life in Alberta.

The oil and gas industry makes up about 50 per cent of the Alberta economy and employs one in six Albertans, with substantial related employment created in sectors such as hospitality, transportation, food services, consultation, construction and real estate. Each dollar invested in the oil and gas sector creates three dollars of value in Alberta's economy across the province, particularly in rural communities.

The next important step is to address the regulatory competitiveness of the province and take steps to improve the regulatory system.

"An improved regulatory framework must maintain environmental standards while improving the efficiency of the system. At the same time the oil and gas industry must continue to improve environmental performance," Collyer said. "We are encouraged by the process the government has established and the ambitious timeline adopted for addressing regulatory reform."

CAPP represents companies, large and small, that explore for, develop and produce natural gas and crude oil throughout Canada. CAPP's member companies produce about 90 per cent of Canada's natural gas and crude oil. CAPP's associate members provide a wide range of services that support the upstream crude oil and natural gas industry. Together CAPP's members and associate members are an important part of a $110-billion-a-year national industry that provides essential energy products. CAPP's mission is to enhance the economic sustainability of the Canadian upstream petroleum industry in a safe and environmentally and socially responsible manner, through constructive engagement and communication with governments, the public and stakeholders in the communities in which we operate."

www.capp.ca

Investment
Each dollar invested in the province's oil and gas industry creates three dollars of value in Alberta's economy. The more attractive our province is for investment, the more Albertans benefit.

Restoring investor confidence does not mean instant prosperity. However, a reputation for strength and stability positions Alberta for positive and long-term economic growth and benefits.

Revenue
In 2008, the petroleum industry invested $54 billion in Canada, including $39 billion in Alberta. Additionally, our industry paid $8.5 billion to the federal and provincial governments in corporate income taxes and $10.7 billion in royalties to Alberta.

When you fill up your car with gasoline or pay your natural gas heating bill, you are the final link in a long chain of businesses that make it possible for us to enjoy these clean, convenient and economical forms of energy. The entire chain is known as the petroleum industry. However, the industry is usually divided into three major components: upstream, midstream and downstream.

The upstream industry finds and produces crude oil and natural gas. The upstream is sometimes known as the exploration and production (E&P) sector. Because Alberta accounts for more than 80 per cent of Canada's oil and gas production, many upstream businesses are based in Alberta and most have their head offices in Calgary.

The upstream petroleum industry in Canada includes more than 1,000 exploration and production companies as well as hundreds of associated service businesses such as seismic and drilling contractors, service rig operators, engineering firms and various scientific, technical, service and supply companies.

Upstream industry revenues totalled $63 billion in 2000, of which 53 per cent came from the sale of crude oil. The remainder was from sales of natural gas, natural gas liquids and sulphur. About half of Canada's oil and gas production is exported to the United States. Canada is self-sufficient in natural gas – supplying virtually all domestic markets with domestically produced natural gas – but imports of crude oil into Eastern Canada account for about 40 per cent of the nation's oil supply.

The Canadian upstream petroleum industry has attained an international reputation for excellence in many areas, including:

- high-tech exploration and production methods;
- cold-climate operations;
- development of oil sands, heavy oil and sour gas;
- gas processing, sulphur extraction and heavy oil upgrading;
- construction and operation of pipelines;
- specialized controls and computer applications;
- services, equipment and training for environmental protection and safety.

Petroleum Services Association of Canada – www.psac.ca

Day crew coming off shift on Nabors rig #51. Left to right: Dave Dylke, Jason Hall, Russel Bazian, Steven Hill, Devin Scherger and Daniel Wray.

Night crew ready to begin their shift on Nabors rig #51. Left to right: Dave Dylke, Nathan Jess, Greg Elliott, Jason Moore, Ron Bailer and Carey Gavel.

Oil workers at Canner rig #1 on the Vermillion oil fields Borradaile, Alberta.
(Canada Science and Technology Museum, CN003640)

Brandon Erdos and Tyler Rook working the drilling floor of Nabors rig #19 near High River.

MIDSTREAM

The midstream industry processes, stores, markets and transports commodities such as crude oil, natural gas, natural gas liquids (NGLs, mainly ethane, propane and butane) and sulphur. The midstream provides the vital link between the far-flung petroleum producing areas and the population centres where most consumers are located. In Canada, transmission pipeline companies are a major part of the midstream petroleum industry. Most of these companies are also based in Calgary, although their activities extend across the country, into the United States and sometimes abroad.

Suncor refinery, Edmonton

Imperial Oil Resource's Strathcona refinery, Edmonton.

Imperial Oil Resource's Strathcona refinery, Edmonton.

DOWNSTREAM

Imagine what John George "Kootenai" Brown would think if he could see how the industry he is credited with starting – by being the first man in recorded history to sell a petroleum product in Alberta – has so changed the province since his own foray into petroleum marketing in 1874.

"The downstream industry includes oil refineries, petrochemical plants, petroleum products distributors, retail outlets and natural gas distribution companies. Although many downstream companies are headquartered in Calgary, the largest centres of activity are near Sarnia, Ontario, and Edmonton, Alberta.

The downstream industry touches every province and territory-wherever consumers are located-and provides thousands of products such as gasoline, diesel, jet fuel, heating oil, asphalt, lubricants, synthetic rubber, plastics, fertilizers, antifreeze, pesticides, pharmaceuticals, natural gas and propane."

Petroleum Services Association of Canada – www.psac.ca

(Glenbow Archives NA-2539-19)

John George "Kootenai" Brown, Pincher Creek, ca. 1899.

Cook's Auto Service, Edmonton,1926. This 15-passenger, Studabaker bus provided service between Edmonton and Lacombe.
(Glenbow Archives ND-3-3350)

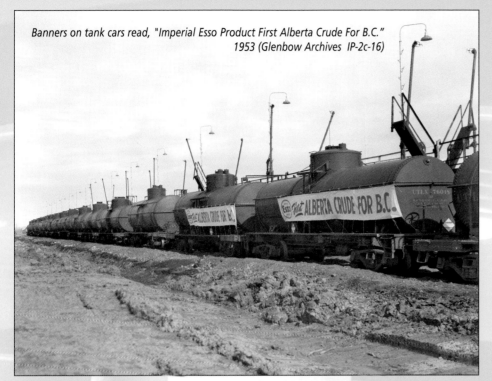

Banners on tank cars read, "Imperial Esso Product First Alberta Crude For B.C." 1953 (Glenbow Archives IP-2c-16)

Imperial Oil tank truck parked beside storage tanks. ca. 1920-1923 (Glenbow Archives IP-2B-13)

Petro-Canada service station, Calgary. In 2008 Suncor bought rival Petro-Canada in a friendly takeover that saw them emerge with a 60% stake in the new Suncor. Petro-Canada will continue on the retail side.

EDMONTON

The development of the City of Edmonton, from Hudson's Bay Company trading post in 1795 to present day metropolis, is a microcosm of the development of this part of Canada as a whole. Today Edmonton is an industrial city, thanks largely to the discovery of oil just south of town in the late 1940s. In the late 1800s, Edmonton was the last supply point for prospectors heading to the Klondike with golden dreams dancing in their heads. For most, those dreams turned to nightmares, but what was a bust for prospectors, became a boom for Edmonton. Many who returned from the Klondike empty-handed, settled in Edmonton and, in ten short years, the city grew to six times its previous size. Because of this the city was made the capital of Alberta in 1905. Its strategic location earned it the reputation as the "gateway to the north", a title that remains to this day.

When you mention Edmonton to most people, they think of the West Edmonton Mall, one of the largest indoor shopping centres in the world. The massive complex contains hundreds of stores, several hotels, a full size hockey rink, amusement park with a very large 'outdoor-sized' roller coaster, a hard-to-believe-until-you-see-it indoor water park, complete with giant, tropical wave pool and water slides, dolphin shows, a pirate cove, themed streets and restaurants galore. This mall is one of the few places where you are likely to drop before you've completed the shop part!

The skyline of the city of Edmonton with the North Saskatchewan River in the foreground.

Alberta Legislative building.

Visitors to this northern Alberta city can't help being amazed at the amount of green space city fathers – and forefathers – have preserved for their citizens. The North Saskatchewan River runs through the heart of the city and at 7,340 hectares or 18,300 acres, Edmonton's river valley parkland is one of the largest areas of urban green space in any North American city.

The turning point for the modern city of Edmonton occurred on February 13th, 1947, some 40 kilometres to the south. After drilling 133 dry wells, Imperial Oil struck it rich with well #134. On that bitter cold day, Vern "Dry Hole" Hunter and his crew changed the fate of Edmonton, and all of Alberta, forever. First spurting water, then drilling mud, and finally oil, Leduc #1 well announced the discovery of the 200 million-barrel Leduc oilfield. In the 25 years following the Leduc oil discovery, the population of Edmonton quadrupled! It was the first in a series of postwar oil and gas finds that forever changed the face of Alberta.

Near the oilfields, existing towns grew with the industry and entirely new towns were built. Virtually overnight, petroleum and petrochemical refining took over Alberta's economy, bringing thousands of new jobs – and job-seekers – to the province. And it all started that cold February day in 1947.

Heart of Edmonton's business district.

Suncor refinery, Edmonton.

IN THE **25** YEARS FOLLOWING THE LEDUC OIL DISCOVERY, THE POPULATION OF EDMONTON QUADRUPLED!

135

EPCOR Gold Bar Wastewater Treatment Plant

Gold Bar's 11 bioreactors employ an intensive biological process known as Biological Nutrient Removal for removing organic pollutants and nutrients (phosphorus and ammonia).

Around the world, water is becoming a more valued resource and is rivaling oil in its importance. In northern Alberta, there is a relative abundance of water, although things could quickly change if sound watershed management practices fail to keep pace with industrial development. Fortunately, these competing demands are working together and resulting in some innovative approaches in watershed management. An award-winning project between, Suncor Energy and water and wastewater management specialists, EPCOR is an example.

EPCOR's Gold Bar Wastewater Treatment Plant and the Suncor Edmonton Refinery (formerly Petro-Canada) gained national attention in 2002 for developing the largest membrane-based water reuse project in Canada. The project involved designing and building a robust membrane treatment facility at the Gold Bar Plant capable of producing a reliable supply of high grade process water for the large oil refinery.

In addition, it linked the two sites with a 5.5 kilometer pipeline – one that would cross the North Saskatchewan River twice, traverse two city parks and a provincial one, and cross two municipal boundaries.

When the partners first came together, Suncor was facing a need for more on-site water to produce hydrogen and steam, and had plans to build a large water treatment facility on or near the refinery to produce high quality water. Rather than directly withdrawing additional water from the North Saskatchewan River, Suncor approached the EPCOR Gold Bar Plant to determine if treated wastewater from the plant could be used instead.

EPCOR started delivering treated wastewater to Suncor in December 2005. Today, up to 15 ML/d is sent to Suncor from the Gold Bar Plant.

Secondary clarifiers produce a clear liquid, some of which will undergo further processing – or "polishing" – in the membrane treatment facility. High-grade process water for industry is the result.

A two foot diameter pipeline crosses the North Saskatchewan River suspended beneath two pedestrian bridges. The 5.5 km pipeline connects the Gold Bar Plant with the Suncor refinery.

The Process

The Gold Bar Plant's membrane treatment facility employs GE hollow fiber membrane technology. By taking what is a relatively small portion of the plant's daily treated volume, it further cleans or 'polishes' the effluent. The membrane pores that filter the effluent are approximately one thousandth the diameter of a human hair. It's filtration taken to a microscopic level.

When it reaches the Suncor refinery, the water is refined even further by a different membrane system employing reverse osmosis. Trace contaminants are removed at a molecular level.

The Gold Bar treated effluent is an ideal water source compared with river water, which is subject to seasonal fluctuations in quality, because it is maintained at a high quality throughout the year. In addition, this reuse solution is more sustainable for the environment as less water is diverted from the river. If Suncor hadn't worked directly with EPCOR on this initiative, Suncor would have had to invest in new equipment required to treat water taken directly from the North Saskatchewan River.

The Facility

EPCOR's Gold Bar Wastewater Treatment Plant is a 310 ML/d tertiary treatment facility, which serves the greater Edmonton area and its more than 780,000 residents. In terms of volume, the daily output of treated wastewater discharged to the North Saskatchewan River would fill an Olympic-size pool more than 110 times a day. So when it comes to supplying Suncor with up to 15MLD -- representing 5 per cent of the plant's daily treatment volume – it's more than well equipped.

This innovative project has the potential to be replicated in the future. In this case, the result was that both companies were able to focus on their core business and what they do best – oil refining for Suncor, and wastewater treatment for EPCOR. EPCOR is actively trying to convert more effluent for water reuse in future by working with industry and local and provincial regulators.

www.epcor.ca

REDWATER DISCOVERY WELL

Wildcat well, Imperial #1, Redwater, 1948. (Glenbow Archives, na-2497-13)

On the morning of August 21, 1948, Christine Cook ran to the farmhouse saying, "The derrick's on fire!" Hilton Cook and son Maurice rushed out and saw oil blowing over the top of the derrick. Redwater Discovery Well, Imperial Redwater #1, was put into production on October 1, 1948. The well flowed at a rate of 1728 barrels per day and Redwater was suddenly on the map! The town grew from a population of 99 in 1947 to approximately 4000 by the end of the decade. Many different companies moved into the area as wells were drilled at a rate of one per day. By 1952 a peak of 926 producing wells pumped out of a 1.3 billion barrels reservoir - 3 1/2 times larger than Leduc. The oilfield revealed by this well is in the Upper Devonian Leduc formation at a depth of 3131 feet. The resevoir is in a fossil coral reef formed 360 million years ago. The field, with an area 15 kilimetres long by six kilometres wide, has an estimated recovery of 700 million barrels of oil. Interestingly, this and similiar resevoirs, offer intriguing possibilities for carbon capture and storage (CSS) in the 21st century. Instead of releasing CO_2 emissions into the atmosphere, CCS involves injecting the CO_2 into secure geological formations such as saline aquifers and depleted reservoirs of coal, oil and gas. See opposite page.

Twice rescued from the scrap heap the original derrick (above) now stands at the first hole of the Redwater Golf course in the heart of the town. The original well is still producing but the site is much less dramatic today as the well pump is underground.

Redwater Discovery Derrick in its position beside the first tee of the Redwater Golf Course.

LEDUC REDWATER REEF

The Leduc Redwater Reef and the underlying Cooking Lake Aquifer together form an integrated system with unparallelled CO_2 storage opportunities. One of the largest of the Leduc reefs, the Redwater Reef is nearly 600 square kilometers in size, more than 1,000 metres deep and up to 275 metres thick. The reef contains the third largest oil reservoir in Canada and experiences a strong water drive from the highly-permeable Cooking Lake aquifer that lies underneath.

While the top of the reef offers potential for CO_2-enhanced oil recovery, the rest of the reef offers a very large capacity for CO_2 storage. "The Leduc Redwater Reef could potentially store as much as one billion tonnes of CO_2," says Dr. Bill Gunter, the Alberta Research Council's principal scientist for carbon capture and storage. "That would allow permanent storage of up to 20 years' worth of (current) cumulative CO_2 emissions from the oil sands." Dr. Gunter calls carbon capture and storage a great 'sunset technology' for Alberta's oil and gas industry as reserves decline. "It would help prolong the life of the industry because more oil and gas could be produced from depleted CO_2 emissions are reduced," he says.

With output from the oil sands expected to double over the next 15 years, the area around Fort McMurray is likely to be the single largest source of growth in Canada's greenhouse gas emissions. However, the Fort McMurray area holds no opportunities for the geological storage of CO_2 because the basin in that corner of the province is too shallow.

The Redwater Reef's access point is located near the Heartland Industrial region, where refineries, petrochemical plants and oil sands upgraders continue to be expanded and it is also one of the closest large capacity sites for storage to Fort McMurray. This positions the reef as an ideal location for early demonstration of carbon capture and storage on a commercial scale.

Courtesy Alberta Research Council
2007-2008 Annual Report

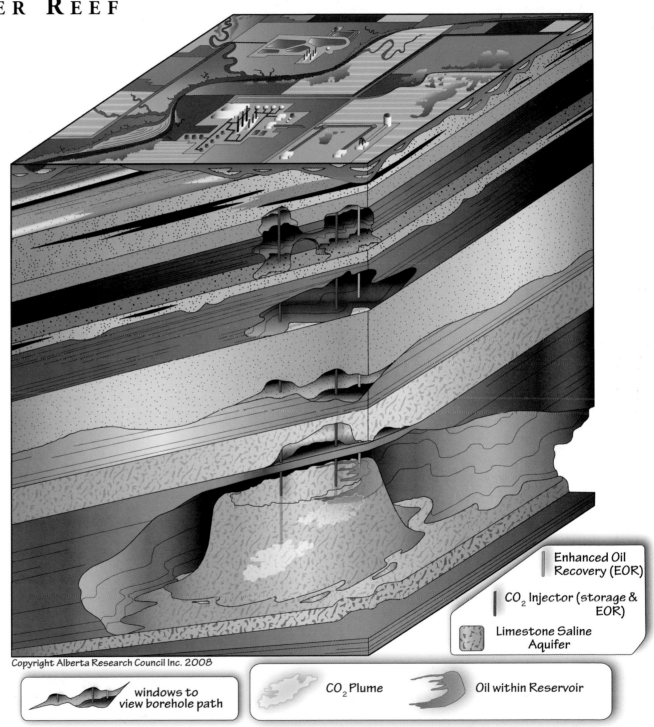

Copyright Alberta Research Council Inc. 2008

Enhanced Oil Recovery (EOR)

CO_2 Injector (storage & EOR)

Limestone Saline Aquifer

windows to view borehole path

CO_2 Plume

Oil within Reservoir

ALBERTA RESEARCH COUNCIL, 1921-2009
NOW PART OF ALBERTA INNOVATES

Alberta Innovates – Technology Futures' offices and labs in Edmonton.

For over 89 years, the Alberta Research Council has collaborated with business, industry, and academia to foster the economic development and prosperity of Alberta. Established in 1921 by a provincial government Order-in-Council as the Scientific and Industrial Research Council of Alberta (SIRCA) with a mandate to carry out research into the potential of the province's natural resources for industrial development, ARC was the first provincial research organization in Canada.

Over the decades, ARC expanded in many directions, adding programs and partnerships and phasing out others in sync with the province's economic objectives. Its history chronicles developments in the province's energy and agriculture sectors.

Tracking provincial advances in transportation and environmental protection, ARC reflected the blend of industrial and scientific resources that made Alberta an internationally recognized centre of high technology innovation and development.

The Alberta Research Council is a story of determination, of innovation, and of unswerving focus on the sustainable development and quality of life for all Albertans. The province has a rich history, full of significant achievements and milestones throughout the decades.

AITF research lab.

Communications advisor, Bonni Clark in the lobby of AITF building, Edmonton.

Each year, ARC served more than 950 customers and partners around the world in the energy, life sciences, agriculture, environment, forestry and manufacturing sectors. ARC also worked with the Alberta government to deliver Alberta's innovation agenda.

On January 1, 2010, ARC became part of the new Alberta Innovates – Technology Futures provincial corporation. AITF builds on the strengths and success of the former ARC, Alberta Ingenuity Fund, iCORE, and nanoAlberta. AITF currently employs over 650 scientists, engineers and professional staff who operate in five facilities located throughout Alberta – in Edmonton, Calgary, Vegreville and Devon. Tech Futures customers have access to leading edge expertise, equipment and facilities. The organization offers a variety of flexible working arrangements to meet our customers' needs including joint ventures, consortia and strategic partnerships.

AITF is a veritable treasure-trove of news reports, feature stories and independent experts for the news media. As an essential link between scientific research and market deployment since 1921, ARC and now AITF helps drive an economy fuelled on brainpower and innovation. Staff members aim to ensure that technology innovations translate into new jobs and industry advancement to create and maintain sustainable prosperity in the province of Alberta.

AITF maintains its role as an independent resource to the community with integrity and pride. Public and private organizations call upon their team of experts, many of whom are world-renown in their areas of study, for their knowledge and science. Tech Futures team members possess a solid perspective on the relevance and impact their work has within the community and the world.

www.albertainnovates.ca

In-Situ

Imperial Oil Resources pumpjacks on cyclic steam-stimulation (CSS) pad near Cold Lake.

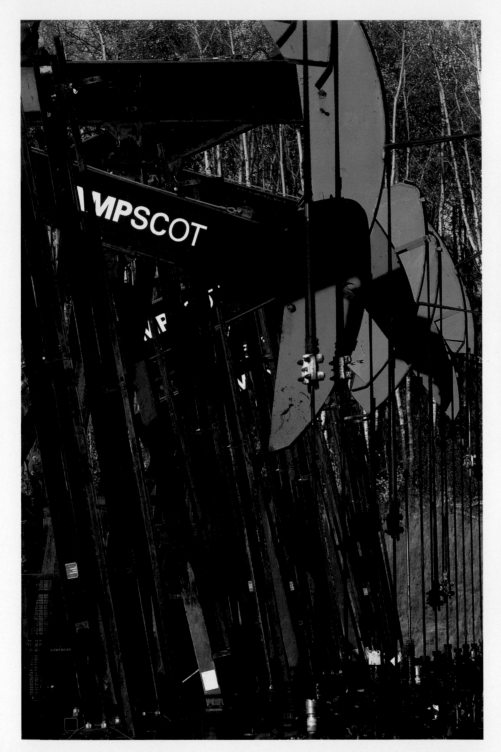

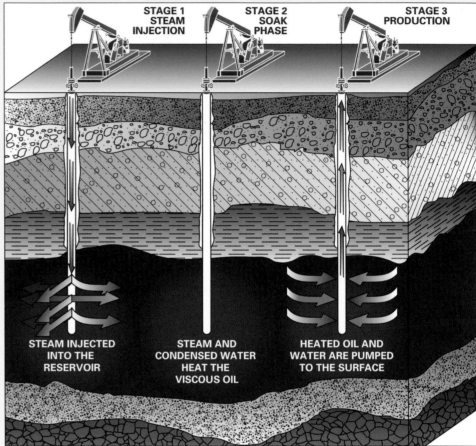

STAGE 1
STEAM
INJECTION

STAGE 2
SOAK
PHASE

STAGE 3
PRODUCTION

STEAM INJECTED
INTO THE
RESERVOIR

STEAM AND
CONDENSED WATER
HEAT THE
VISCOUS OIL

HEATED OIL AND
WATER ARE PUMPED
TO THE SURFACE

The in-situ recovery method at Cold Lake uses a process called cyclic steam-stimulation.
(Diagram courtesy of Imperial Oil Resources)
Two of Imperial Oil's production pads at their Cold Lake Operations (below and left).

CYCLIC STEAM-STIMULATION

Mahkeses plant at Imperial Oil's Cold Lake Operations. The large vessel on the right of the photo is the Inlet Flow Splitter which separates the inlet production stream (from the production pads) into oil/water and gas.

Terry Pelechosky controls and monitors every aspect of the plant's operations from the control room at the Mahkeses plant.

The *in-situ* recovery method at Cold Lake uses a process called cyclic steam-stimulation (see diagram opposite). Steam is injected at high temperature and pressure through wells into the oil sands to thin the heavy bitumen and enable it to flow. Surface-based pumps lift the heated water-bitumen mixture through the same well to the surface for separation and processing. The bitumen is then blended with a lighter hydrocarbon liquid and shipped to market by pipeline.

In addition to plant and field facilities for steam generation and bitumen production, the Mahkeses project includes an electrical co-generation plant. The plant is built around two 85-megawatt, natural-gas-fired turbines. The turbines generate electricity, while heat-recovery steam-generators capture exhaust heat from the turbines and use this heat to generate steam. Co-generation results in higher overall energym efficiency and lower overall costs by using this waste heat from the turbine generators to produce steam. About 60 percent of the electricity generated by the facility is currently being used for Imperial's Cold Lake operation, with the balance entering Alberta's power pool.

STEAM-ASSISTED GRAVITY DRAINAGE - SAGD

Aerial view of Suncor's MacKay River SAGD facility.

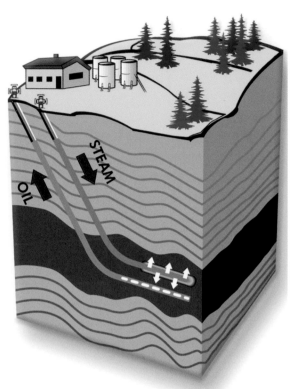

Diagram showing the steam-assisted gravity drainage (SAGD) method of recovering bitumen from reservoirs too deeply situated for surface mining methods.

Diagram and photo courtesy: Suncor

About 80% of the oil sands in Alberta are buried too deep below the surface for open pit mining. Suncor is an *in-situ* player and produces bitumen at its MacKay River facility using an *in-situ* technology called steam-assisted gravity drainage (SAGD). Surface mining methods using large trucks and shovels – like those at Syncrude – are not practical at a facility like MacKay River because the oil sands deposits are too far underground. So instead, SAGD combines horizontal drilling with thermal steam injection. A pair of wells is drilled into the ground about five metres apart and steam is injected into the reservoir through the top well. The steam softens the bitumen and enables it to flow out of the reservoir and into the lower well. From there, it's produced to the surface.

One way to think of SAGD is "heating the oil and catching the drips."

The well pairs, drilled from a central "well pad" are capable of producing approximately 1,200 barrels of bitumen daily for about eight years. After that, the well pads are reclaimed with original topsoil kept specifically for that purpose. The underground bitumen recovery means the environmental surface impacts are minimal.

More than 90 per cent of water used to generate steam at MacKay River is recycled, a key feature of the environmental efficiency of the operation. Features such as electricity cogeneration, 90-per-cent water recycling and minimal surface impact highlight our commitment to responsible development and reducing emissions.

Challenges facing *in-situ* process are efficient recoveries, management of water used to make steam, and co-generation of all (otherwise waste) heat sources to minimize energy costs. Other methods of *in-situ* recovery look promising, and are in development.

CO-GENERATION

Energy recycling is the energy recovery process of utilizing energy that would normally be wasted, usually by converting it into electricity or thermal energy. Undertaken at manufacturing facilities, power plants, and large institutions such as hospitals and universities, it significantly increases efficiency, thereby reducing energy costs and greenhouse gas pollution simultaneously. The process is noted for its potential to mitigate global warming profitably. This work is usually done in the form of combined heat and power (also called co-generation) or waste heat recovery.

The cogeneration plant at MacKay River is an integral part of Suncor's bitumen extraction facilities. The 165 megawatt (MW) natural gas-fired cogeneration power plant provides ten MW of power to the MacKay River site and an additional 60 MW will be sold under long-term contracts. Surplus power of approximately 95 MW will be supplied to the Power Pool of Alberta.

These four generators produce the steam that is used to release the bitumen from underground reservoirs so that it can be brought to the surface.

Cable tool rig on display at the Leduc #1 Energy Discovery Centre, Devon. This is the other end of the technology spectrum as reflected at left.

Even Canadians don't comprehend what they're sitting on

Invention of horizontal drilling opened up global bonanza in oil sands

The following was first published in the *Toronto Globe & Mail* in May of 2010. Neil Reynolds

Ottawa historian and storyteller Alastair Sweeny calls Roger Butler "one of the true benefactors of humanity," a Canadian inventor who ranks with such famous innovators as Abraham Gesner (the geologist who invented kerosene in 1854 and "lit up the world for 50 years"), Charles Saunders (the botanist who developed frost-resistant wheat in 1892 and multiplied global crop production), Reginald Fessenden (the broadcast pioneer who, in the early 1900s, may have been the first person to transmit the human voice by radio wave) and which led eventually to Mike Lazaridis giving the world the BlackBerry in the late 1990s.

Mr. Sweeny celebrated Mr. Lazaridis last year in *BlackBerry Planet: The Story of Research in Motion and the Little Device That Took the World by Storm*, his fourth "business biography." In *BlackBerry Planet*, Mr. Sweeny calculated the economic and cultural consequences of "the little device" as enormous, taking a huge step toward what he calls "the TeleBrain," a powerful, portable micro-computer with the revolutionary capability of vastly increasing humanity's brain-power in the next 20 years.

Now, in *Black Bonanza: Canada's Oil Sands and the Race to Secure North America's Energy Future*, his fifth "business biography," Mr. Sweeny calculates the economic and political consequences of Roger Butler's discovery of SAGD (steam-assisted gravity drainage), the oil-sands process that made Canada an energy superpower – with more than one trillion barrels of recoverable oil, with another two trillion barrels in reserve.

Canadians now know these numbers, Mr. Sweeny says, though few Canadians appear to recognize the global importance of them.

"If energy is the master resource of the human race, then Canada is truly blessed," he writes. "Beneath the boreal forests of Alberta and Saskatchewan, halfway between Edmonton and the border of the Northwest Territories, lies a black bonanza of oil-soaked sand.

"It's hard for people to grasp the simple fact – that [these sands comprise] the largest known petroleum assets on the planet," he writes. "Covering an area larger than England, this belt of oil-soaked silicon dwarfs the light oil reserves of the entire Middle East."

Even oil industry analysts don't get it, Mr. Sweeny says, citing authorities who still rank Canadian oil reserves as #2, behind Saudi Arabia. "This fiction persists in the face of the evidence that the Athabasca sands are far larger," he says. "A trillion barrels of synthetic crude is four times larger than Saudi Arabia's 250 billion barrels." He notes that the International Energy Agency persists in listing Canada's oil reserves at a mere 175 billion barrels.

Mr. Sweeny asserts that the oil sands now hold irreversible advantages over offshore drilling, presciently comparing BP's difficulties – in drilling an offshore Gulf of Mexico well "that goes as deep as Mount Everest goes high" – with oil sands operations.

BP is now dealing with an environmental disaster of perhaps historic dimensions, inadvertently increasing the U.S. need for more Canadian oil sands production.

"The sands are just lying there for the taking, some of them up to 140 feet thick," Mr. Sweeny says. "All you have to do is build a giant washing machine, pay a royalty to a friendly government and promise to clean up when you leave."

If the oil sands are a "blessing" for Canada, Mr. Sweeny asserts, they are also a blessing for the world. The oil sands ensure a stable supply of oil to the United States, which will become less and less reliant on imports from "dictatorial regimes" that seek to hold the world to ransom. Thus, had he still been alive, Roger Butler might well qualify for a Nobel Peace Prize for his invention of horizontal drilling.

Although not widely recognized by Canadians, Butler's name is still spoken in reverence in the engineering community, Mr. Sweeny says. Butler himself, he says, was always modest about his discovery. In fact, though, Butler (who died five years ago) "gave the citizens of the planet a hundred years or so of energy security that we never thought we had." A true benefactor indeed.

Mr. Sweeny's Black Bonanza is more than rah-rah. He critiques a wealthy industry that has failed to fulfill its environmental obligations – and he argues that it must be required to do so. But he regards the environmentalists' doomsday assault on the oil sands as preposterous: "I agree totally that we must . . . clean up the planet. But I believe [we can do it] without obsessing about a trace gas that helps plants grow." Goats have done more damage to the environment, he suggests, than carbon dioxide.

"THE SANDS ARE JUST LYING THERE FOR THE TAKING, SOME OF THEM UP TO **140** FEET THICK
ALL YOU HAVE TO DO IS BUILD A GIANT WASHING MACHINE, PAY A ROYALTY TO A FRIENDLY GOVERNMENT
AND PROMISE TO CLEAN UP WHEN YOU LEAVE."

Syncrude upgrader, Mildred Lake facility, Fort McMurray.

"When the historian in me looks at the history of climate change," he says, "he notes that we have significant periods of naturally occurring heating and cooling." A thousand years ago, Britain was covered in vineyards – and Viking farmers tended cows in Greenland. Four hundred years ago, the sea froze between England and France. Mr. Sweeny, for all his scholarly analysis, expresses his doubts bluntly in the vernacular: "I smell a rat."

Special to The Globe and Mail

THE ATHABASCA

The total annual allocation of water from the Athabasca River
for all uses (e.g. municipal, industrial and oil sands)
is less than 3.2 per cent of flow.
Current oil sands mining projects use about one per cent
of the total annual water flow of the Athabasca River.

The Athabasca River at Fort McMurray in the golden light before sunrise.

The Canadian boreal forest covers two million square miles.
The oil sands mines have so far impacted about 150 square miles
or one hundredth of one percent of the total area.

The Athabasca River north of Fort McMurray.

FORT McMURRAY

To better accommodate oversized truck loads, a new bridge (above right) is being built across the Athabasca River. Located adjacent to the two existing bridges, the new bridge will be five lanes wide and will be dedicated to northbound traffic. It is scheduled to open in 2011.

www.oilsandsdevelopers.ca

Passengers boarding Canadian North flight in Edmonton destined for the oil sands.

Baggage being loaded onto an Enerjet 737 in Edmonton destined for the oil sands.

In 2008, the population of Fort McMurray and the communities that make up the Regional Municipality of Wood Buffalo exceeded 100,000, a growth of 17 per cent from the previous year. This explosive growth and the distance to Edmonton have created many challenges in housing, municipal services and transportation. While the housing in Fort McMurray resembles that of any Canadian community, it can't begin to meet the needs of all workers and one of the solutions has been camps and lodges.

PTI maintains the largest chain of permanent base camps in the industry, strategically located in remote areas where many companies need to house staff at a common location. From shared rooms to private suites, the company offers a complete range of accommodation options to house staff for any length of time. For the last few years camps and lodges like PTI's have hosted well over 20,000 mobile workers throughout the oil sands region.

Highway 63 connects Fort McMurray to Edmonton, 430 kilometres (270 miles) to the south. It sees a lot of heavy truck traffic and is considered by many to be one of the more dangerous roads in Alberta. After a particularly tragic bus accident in 2007 where five workers were killed, many oil sands developers instituted air transportation of their own, investing millions on private airstrips to provide workers with safer and easier access to remote work sites and to Edmonton.

The Fort McMurray Airport oversees the single busiest airstrip of any small or medium-sized city in Canada. More than 500,000 passenger movements and 70,000 take-offs and landings occurred at the airport in 2009.

Typical housing in Fort McMurray suburb of Timberlea.

PTI's Athabasca Lodge near Fort McKay has 1550 rooms.

Shell Canada's Bombardier RJ taking off with a load of oil workers headed back to the job.

Highway 63 carries the highest tonnage per kilometre in the country and accommodates the largest and heaviest loads ever carried on highways anywhere in the world.

Fort McMurray is an urban service area within the Regional Municipality (R.M.) of Wood Buffalo, Alberta. Canada's energy capital is home to the Athabasca Oilsands which contain enough oil reserves to maintain production for generations to come. The community has played a significant role in the history of the petroleum industry in Canada. Oil exploration is known to have occurred early in the 20th century but Fort McMurray's population remained very small, no more than a few hundred people. By 1921 there was serious interest in developing a refining plant to separate the oil from the sands. Alcan Oil Company was the first to begin bulk tests at Fort McMurray.

In 1967, the Great Canadian Oil Sands (now Suncor) plant opened and Fort McMurray's growth skyrocked. More oil sands plants followed, particularly after 1973 and 1979, when serious political tensions and conflicts in the Middle East triggered oil price spikes. The population of the town reached 6,743 by 1971 and climbed swiftly to 30,772 by 1981, a year after its incorporation as a city. It is well over 60,000 today and showing no signs of slowing down. With over $125 billion in oil sands investment planned for the next decade, Wood Buffalo is indisputably the world's hottest regional economy.

The Franklin Hotel and a drugstore on Main Street, Fort McMurray,.early 20th century.. (Glenbow Archives, NA-670-48)

View to the north up Franklin Avenue in the winter of 1929. (Glenbow Archives, NA-1258-32)

Imagine a traveller arriving at the Franklin Hotel (far left) early in the 20th century. He would probably have revelled in the luxury and amenities he would enjoy inside – particularly compared to life on the trap-line or in whatever line of work drew him so far north in his time. Now imagine this same gent – with the benefit of time-travel – checking in to one of the suites in the contemporary iteration of the original Franklin (right). Flat-screen television, wi-fi, microwave, coffee-maker, telephone, hot shower, indoor toilet, air-contitioning, electric lights and a pair of elevators to whisk him and his pack down to the lobby in the morning.

Franklin Suite Hotel.

View to the north up Franklin Avenue today reflects just over 80 years of growth from the image at far left.

St. Jean Baptiste Catholic Mission.   Photo: Tammy Plowman

Walter Hill's Drugstore.   Photo: Tammy Plowman

FORT MCMURRAY HERITAGE PARK

Heritage Park is Fort McMurray's own little village locked in time. Come to the park during the summer months and take a stroll through history. No appointment needed!

On your self-guided tour you will find historic buildings and artifacts which showcase this community's unique history. From the early 1900s trapper's cabin and Catholic Mission to Walter Hill's Drugstore from the 1930s, Heritage Park preserves Fort McMurray and region's past for generations to come. Guided tours of Heritage Park are also available.

FORT MCMURRAY MARINE PARK

The historic vessels were integral to river transportation along the Athabasca and Clearwater Rivers and as far north as the Arctic Circle. The ships, owned by the Society will serve as the primary artifacts and feature icons of Fort McMurray Marine Park.

The Park will be located along the Clearwater River in the Fort McMurray lower town site. The four-acre site is the original location of the shipyards where after spring break up, passengers and freight destined for the north were transferred from rail to ship or barge. It is the only shipyard remaining in the province of Alberta and its significance in Alberta's history awaits unveiling for the enjoyment of residents and visitors.

www.fortmcmurrayhistory.com

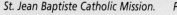

Hudson'a Bay Company steamer "Grahame" at Fort McMurray, ca. 1899-1900.   (Glenbow Archives, NA-4035-98)

Canadian Coast Guard Ship (CCGS) Miskanaw, McMurray tug boat, barge and crane on back of CCGS Dredge 250 at the future site of the Fort McMurray Marine Park.

OIL SANDS DISCOVERY CENTRE

The Oil Sands Discovery Centre in Fort McMurray is in the heart of the world's biggest single oil deposit – Alberta's Oil Sands. At the Centre you'll be surrounded by BIG things – a dragline bucket, a 150-tonne heavy hauler with tires three metres high and "Cyrus," an 850-tonne bucketwheel excavator. You'll really get the "big picture" when you see our big screen movie "Quest for Energy". Discover us soon!

www.oilsandsdiscovery.com

Aerial view of the Oil Sands Discovery Centre.

Wabco 150, 150-tonne oil sands truck used in the Suncor mine.

Using oil sand mined near Fort McMurray and trucked to Edmonton, the plant extracted bitumen from up to 15 tonnes of oil sand an hour. Each piece of equipment in this pilot plant was based on the mechanical and process design of the planned Syncrude plant in Fort McMurray, but at a scale of 1:200.

Pilot plant operations helped engineers and scientists fine-tune the extraction process and work out many of the significant technical issues in scaling up to the Syncrude plant that opened in 1978. It represents a major step in the commercialization of oil sands production in the early 1970s and is a rare surviving eample of an engineering prototype that led to the development of a multi-billion dollar Industry.

Pilot extraction plant built in 1971 for Syncrude to test the commercial feasibility of its proposed extraction process.

The Alberta government estimates that the province's three main oil sands deposits
contain 173 billion barrels of oil that are economically recoverable today.
The Alberta Energy Resources and Conservation Board estimates that more than 300 billion barrels
may one day be recoverable from the oil sands and
it puts the total size of the deposit at 1.7 trillion barrels.

OIL SANDS

Busy day in Shell Canada's Muskeg River Mine.

Aerial view of provincial government's pilot plant for extracting oil from the northern Alberta oil sands, Bitumont., ca. 1949-1950. (Glenbow Archives PA-1599-451-2)
The long-abandoned site still occupies its place of honour (inset right) along the banks of the Athabasca near Shell Canada's Muskeg River mine north of Fort McMurray.

OIL SANDS STORY

Courtesy: Oil Sands Discovery Centre, Fort McMurray

THE RESOURCE: Alberta's oil sands contain the biggest known reserve of oil in the world. An estimated 175 million barrels of recoverable oil are trapped in a complex mixture of sand, water and clay. The most prominent theory of how this vast resource was formed suggests that light crude oil from southern Alberta migrated north and east with the same pressures that formed the Rocky Mountains. Over time, the actions of water and bacteria transformed the light crude into bitumen, a much heavier, carbon rich, and extremely viscous oil. The percentage of bitumen in oil sand can range from 1 per cent -20 per cent. The oil saturated sand deposits left over from ancient rivers in three main areas, Peace River, Cold Lake and Athabasca. The Athabasca area is the largest and closest to the surface, accounting for the large-scale oil sands development around Fort McMurray.

Geologists, surveyors and mine engineers play a considerable role in the mine planning process before any heavy equipment is introduced. The mine plan must commit to return the area to its former environmental condition. G.P.S. is used extensively to pinpoint mining areas.

MINING: Since the 1920s, open pit mining has been central to oil sands development. Mine equipment from the early years was scaled up significantly when large commercial operations started to come on line. The first large scale commercial operation, Great Canadian Oil Sands (now Suncor Energy), introduced German manufacturerd O&K bucketwheels from the coal mining industry when they opened in 1967. Syncrude Canada Limited opened in 1978 and introduced gigantic draglines 60 times as large as the bucket on display from Bitumount, the first commercial oil sands plant. These large machines were connected to the processing plant by a system of conveyor-belts. Today, large trucks and shovels have replaced draglines and bucketwheels as a more selective, and cost effective way to mine oil sands. The process begins by clearing trees, draining and storing the overburden and then removing this top layer of earth to expose the ore body. The equipment must be durable and strong enough to withstand extreme climate and abrasive oil sand. Mining never stops, the trucks and other equipment work day and night, every day of the year. Planning is an essential and continuous part of the process.

The shovel scoops up the oil sand and dumps it into a heavy hauler truck. The heavy hauler truck takes the oil sand to a conveyor belt which transports the oil sand from the mine to the extraction plant. Presently, there are extensive conveyor belt systems that transport the mined oil sand from the recovery site to the extraction plant. With the development of new technologies these conveyors are being phased out and replaced with hydrotransport technology. Hydrotransport is a combination of ore transport and preliminary extraction. After the bitumous sands have been recovered using the truck and shovel method, it is mixed with water and caustic soda to form a slurry and is pumped along a pipeline to the extraction plant. The extraction process thus begins with the mixing of the water and agitation needed to initiate bitumen separation from the sand and clay.

Loaded Caterpiller truck heading for conveyor belt,
Shell Canada's Muskeg River Mine north of Fort McMurray.

EXTRACTION: Dr. Karl Clark, a scientist working for the Alberta Research Council, developed and patented the hot-water extraction technique. Building on earlier experimentation by Sidney Ells and others which used hot water to separate oil from oil sands, Dr. Clark's work brought the process to a commercial scale. Oil sand is mixed with hot water creating a slurry. Early methods used large tumbler drums to condition the slurry. Today, hydrotransport pipelines are used to condition and transport the oil sand from the mine to the extraction plant. The slurry is fed into a separation vessel where it separates into three layers - sand, water and bitumen. The bitumen is then skimmed off the top to be cleaned and processed further. Secondary recoveries are made with the middlings zone of the separation vessels to return the smaller quantities of bitumen that would otherwise flow to the settling ponds. Ph levels and temperature are key variables in the process.

IN-SITU: About 80 per cent of the oil sands in Alberta are buried too deep below the surface for open-pit mining. This oil must be recovered by *in-situ* techniques. Using drilling technology, steam is injected into the deposit to heat the oil sand lowering the viscosity of the bitumen. The hot bitumen migrates towards producing wells, bringing it to the surface, while the sand is left in place ("*in-situ*" is Latin for "in place"). Steam Assisted Gravity Drainage (SAGD) is a type of *in-situ* technology that uses innovation in horizontal drilling to produce bitumen. *in-situ* technology is expensive and requires certain conditions like a nearby water source. Production from *in-situ* already rivals open-pit mining and in the future may well replace mining as the main source of bitumen production from the oil sands.

Challenges facing *in-situ* process are efficient recoveries, management of water used to make steam, and co-generation of all (otherwise waste) heat sources to minimize energy costs. Other methods of *in-situ* recovery look promising, and are in development.

Visitor examining oil seepage from area of recent ecavation, Shell's Muskeg River Mine north of Fort McMurray.

Drag Line and bucket wheel rigs on display at Syncrude, north of Fort McMurray. The display area is called "The Giants of Mining".

UPGRADING: The oil in oil sand is called bitumen, a complex hydrocarbon made up of a long chain of molecules. In order for bitumen to be processed in refineries, this chain must be broken up and reorganized. Unlike smaller hydrocarbon molecules, bitumen is carbon-rich and hydrogen-poor. Upgrading means removing some carbon while adding additional hydrogen to make more valuable hydrocarbon products. This is done using four main processes: coking removes carbon and breaks large bitumen molecules into smaller parts, distillation sorts mixtures of hydrocarbon molecules into their components, catalytic conversions help transform hydrocarbons into more valuable forms and hydrotreating is used to help remove sulphur and nitrogen and add hydrogen to molecules. The end product is synthetic crude oil, which is shipped by underground pipelines to refineries across North America to be refined further into jet fuels, gasoline and other petroleum products.

It must be noted that some of the oil companies pipe their bitumen south in diluted form for upgrading at other refineries. Others produce either a single high-quality synthetic crude oil or multiple petroleum products to suit market feedstock demand.

www.oilsandsdiscovery.com

Shell's Muskeg River Mine north of Fort McMurray.

Drag Lines and bucket wheels were the first machines used by Syncrude to take oil sands from the ground.
(Province of Alberta Archives J.4248.12

163

Muskeg River Mine sits on Shell's Lease 13 which contains more than five billion barrels of mineable bitumen – an amount that's about twice the conventional oil reserves remaining in Alberta. On Lease 13 the oil sands deposit is close to the surface and contains a high concentration of oil, making it ideally suited to mining. As currently designed, the mine will recover 1.6 billion barrels of bitumen. Shell also has approval for Muskeg River Mine Expansion and Jackpine Mine, which will eventually develop most of Lease 13.

As a whole, Canada's oil sands industry currently produces close to 60 per cent of the nation's petroleum needs and has the potential to account for 65 per cent of Western Canadian crude production by 2010. Alone, the Muskeg River Mine supplies almost 10 per cent of Canada's oil needs.

The mine is part of the Athabasca Oil Sands Project, a joint venture among Shell Canada (60 per cent), Chevron Canada Limited (20 per cent) and Marathon Oil Sands L.P. (20 per cent).

Shell Canada's ugrader at their Muskeg River Mine north of Fort McMurray.

Ryan Lemay's six-foot frame illustrates the size of this monster Caterpiller truck at Syncrude Canada.

Interior of Shell's truck maintenance facility at Syncrude's Mildred Lake mine near Fort McMurray.

Trucks dumping oil sand onto conveyor belt at Shell's Muskeg River Mine north of Fort McMurray.

"On a well-to-wheel (lifecycle) basis, the carbon intensity of oil sands-based fuels falls within the range of carbon intensities for other conventional crude-based fuels used in the United States. A well-to-wheel CO_2e analysis calculates the total CO_2e emissions from the production and distribution of the feedstock and the fuel, and from the use of the fuel in the vehicle."

Cambridge Energy Research Associates, *Growth in Oil Sands – Finding the New Balance.*

An empire from a tub of goo

How did the quest to retrieve the treasure hidden beneath huge swaths
of northern Alberta go from fool's errand to monumental payoff?

Erin Anderssen, Shawn McCarthy and Eric Reguly explain.

The following feature article was originally published in the

Murray Smith remembers what happened on the morning of April 9, 2003, the way other Canadians remember Paul Henderson's miracle goal against the Russians. For Mr. Smith, then Alberta's energy minister, the big score was a letter from his federal counterpart south of the border. It was about the oil sands – a resource that had long been underestimated at home and almost ignored internationally. No more, U.S. energy secretary Spencer Abraham wrote. From now on, when the Americans talked oil, they would be counting the reserves sitting beneath the forests of northern Alberta.

Mr. Smith had grown up among the oil rigs of central Alberta and bought his first share in an oil company when he was 11 by collecting his older brother's beer bottles. He had also spent much of his adult life in the oil patch and understood more than most the significance of Mr. Abraham's message. The endorsement from the world's hungriest oil consumer was like winning an Oscar. Keen to reduce its dependence on the Middle East, the U.S. was officially acknowledging for the first time that the tarry mud around Fort McMurray could be turned into gasoline, diesel and heating fuel at a profit.

The world finally was acknowledging what Albertans had been saying for decades: that their oil sands rival any source of crude on Earth. "If you took all the oil in the south of the United States and all the oil in Alaska and all the oil in Mexico," Mr. Smith points out, "it doesn't hold a candle to Alberta."

With rising prices and prospects of a Mideast war prompting concerns about the security of the U.S. supply, media giants from CBS's 60 Minutes and The New York Times flocked to the tale of an oil bonanza so close to home. Enthusiasts outnumbered the skeptics and the phrase "second only to Saudi Arabia" went from speculation to conventional wisdom. Alberta had become a bankable star in the global oil game.

Or as Mr. Smith jokes: "It only took 40 years to become an overnight success."

A decade earlier, the oil sands had resembled a massive boondoggle, backed by only a few believers who struggled to attract capital for faltering projects. And now the race to profit off that pile of dirt spread across an area the size of Florida is transforming the country.

Now, oil production in northern Alberta is expected to quadruple to more than four million barrels a day by about 2020, if all the projects proposed go ahead. Virtually every major oil company in the Western world has picked up a piece of the action, investing nearly $90-billion to create what promises to be the biggest industrial project on Earth and sparking predictions that Canada will become what Prime Minister Stephen Harper calls an "energy superpower."

The oil sands are seen as a crucial source in a world of increasingly tight supply, where many reserves are in politically volatile regions controlled by undemocratic states. Put another way: Should they disappear tomorrow, one industry expert estimates, the price of oil could jump a third to $130 a barrel.

The value and importance of the oil sands will make that much harder the choices that Albertans and all Canadians suddenly face. Canada has now become a major-league merchant of one of the most desirable – and dirtiest – sources of energy. The money is flowing in, and the profits are rolling out – good news for stockholders, the Canadian dollar and government coffers.

But there are environmental and social costs to stuffing our pockets while the oil speeds south. And Canadians will have to answer a question already being asked by many Albertans: When does a boom become a burden?

Syncrude upgrader, Fort McMurray.

The wrinkles are beginning to show. The growing wealth of Alberta has aggravated the cleavage between Central Canadian assumptions and Prairie assertions, between haves and have-nots. The soaring Canadian dollar – viewed more and more as petro-currency – has savaged the Central Canadian manufacturing and forestry industries. The economy is increasingly concentrated on oil, which can be a fickle commodity. As people flock to the province, escalating housing costs have squeezed everyday Albertans and overburdened public services, most acutely in Fort McMurray, where men and women come to work but not to settle.

And it is far from certain that the bullish assumptions about development will pan out. There are at least two major hurdles: the growing protest at home and abroad over the massive environmental toll and a serious shortage of workers to build all those multibillion-dollar projects.

Before beginning to discuss how best to manage this mixed blessing, Canadians may wonder just how they created an empire from a tub of goo.

Oil does not sprout like geysers from wells in northern Alberta. It is trapped in the mud, in the form of bitumen, a thick, pasty hydrocarbon that native people once used to seal their canoes. (The Bible says Noah did the same to waterproof the Ark.)

Early explorers predicted that one day the sands would prove useful. In 1789, Alexander Mackenzie described how the banks of the Athabasca seeped oil into the river. A century later, a member of a delegation that had come to negotiate a land treaty wrote: "That this region is stored with a substance of great economic value is beyond all doubt, and when the hour of development comes it will, I believe, prove to be one of the wonders of Northern Canada."

The hour of development took more than a century to arrive. A tour of the oil sands in the early 1990s would have found only two companies trying to make a go of it and close to wishing they had not bothered. Suncor Energy Inc., partly owned by the Ontario government, had started first, along a stretch of the Athabasca River in 1967. "No other event in Canada's centennial year is more important," Alberta premier Ernest Manning declared on opening day.

But despite all the cheerleading, the project was "a leap of faith," says Paul Chastko, a University of Calgary historian and the author of Developing Alberta's Oil Sands. "It was based on this calculation that we don't necessarily need this resource today, but we may need it in some deep, dark distant day in the future."

A tantalizing treasure lay in wait underground; digging it up with commercially untested methods and still managing to turn a profit was the challenge. And by 1990, after decades of technical hiccups and economic volatility, opening-day optimism had long since flagged. The Suncor mine was plagued by fires and machinery that constantly broke down in cold weather. The situation wasn't much better down the river at Syncrude Canada Ltd., a consortium of U.S. and Canadian companies that had started producing oil in 1978.

Crude prices had recovered from the disastrous lows of the mid-1980s, but were still hovering around $20 (U.S.) a barrel. The cost of production at both operations was not much below that. It was understood, one Suncor former executive recalls, that the company might have walked away then and there, were it not for the billions spent in capital costs and the mammoth environmental liability it had created, most of which was waste water sitting in vast ponds created next to the Athabasca River.

Athabasca River at Fort McMurray.

Rick George recalls his first trip to Fort McMurray as the new chief executive officer of Suncor, which The Globe and Mail had dubbed "the unluckiest company in Canada." It was 1991, and Mr. George, an American who had come to Canada from the booming North Sea oil industry, found himself running an operation dogged by negative politicking and lumbering technology.

"I can remember coming over the horizon and looking at that plant for the first time, and I did wonder what I had got myself into," he says. "But I'm a very optimistic person. I thought we could make a difference and the people here have executed way beyond anything I ever imagined would happen."

It was no easy feat. Unlike the vast majority of the world's bitumen reserves, the first sites being developed around the Athabasca are shallow enough to mine from the surface, although this still requires a great deal of energy and labour.

To put the situation in context, Alberta writer David Finch, the author of Pumped: Everyone's Guide to the Oil Patch, suggests this experiment: "Take molasses out of your kitchen cupboard, put as much sand in there as molasses, stir it up, and then put it outside where it gets cold and thick and won't flow – well, that's what the tar sand is like. It's extremely hard to work with, and it wrecks all your equipment."

This Caterpillar 797B is 7.6 metres (24') high, weighs 623,690 kilos (1,375,000 lbs) empty, takes 6814 litres (1800 gallons) of fuel, carries a whopping 345 tons and costs almost $6 million..

The muddy dirt clogs gears and conveyors; the sand corrodes pipelines. To mine bitumen, the land must first be cleared and drained. Clumps of sand are shovelled out and then mixed with water and heated, to force the hydrocarbon to rise to the top. It is then processed in an "upgrader" to produce synthetic crude before being sent to a refinery and turned into gasoline and heating oil.

Estimates vary, but environmental groups say it now takes two to four barrels of fresh water from the Athabasca plus 750 cubic feet of natural gas and about two tons of oily sand to produce one barrel of oil. The process produces two to three times the carbon emissions of a conventional oil well and creates toxic waste water, called tailings, that cannot be allowed back in the river.

To expand profitably, the companies needed to have two things happen: The technology had to improve and the price of oil had to go up. In the early 1990s, a simple switch helped to solve the first problem. Companies went back to using enormous dump trucks – as big as a two-storey house – to haul the sand, rather than conveyer belts, which were difficult to move when needed and often froze in the northern cold. At Suncor, this yielded immediate results: Energy requirements were reduced by 40 per cent and the overall cost per barrel was slashed by several dollars.

Syncrude also fared better after it figured out how to transport the bitumen more cheaply by sending it through pipes as a watery slurry. Even with the price of oil bouncing around $20 a barrel, the improvements were enough for both companies to turn a profit. But to the rest of the world, the oil sands might as well have been on another planet. The year Mr. George arrived at Suncor, they were producing about 350,000 barrels a day, a tiny fraction of what geologists believe the sands hold.

And how much oil is there? Estimates bounced around for years until 1999, when Alberta got serious about determining its potential. Based on data from 56,000 wells and 6,000 core samples, the Energy and Utilities Board (EUB) came up with an astonishing figure: The amount of oil that could be recovered with existing technology totalled 175 billion barrels, enough to cover U.S. consumption for more than 50 years. With the new math, Canada slipped quietly into second place behind Saudi Arabia's 265 billion barrels in oil reserves, followed by Iran and Iraq.

To the frustration of Albertans, nobody paid much attention. There was no war on terror and the world was awash in oil. The news "went virtually unnoticed," recalls Rick Marsh, a geologist who leads the EUB's oil-sands section.

Visitors to Shell's Muskeg River Mine (right) marvel at the size of this Bucyrus 495HF Electric Rope Shovel. The beast has a gross working weight of 1,315,000 kg (2,900,000 lbs), is 20.72 metres (68 feet) high and costs $15 million. You can just make out the operator in her place in the cab at top-left in the photograph.

Aerial view of giant Caterpiller trucks hauling oil sands to conveyor at belt that will move it out of the mine pit to the upgrader for processing.

Then, in the spring of 2002, Murray Smith, recently installed as Alberta's energy minister, was called to a Saturday-morning meeting to review the EUB's annual report. He spotted the big figure and "his eyes lit up," says Neil McCrank, then the board's chairman. "Murray is a salesman and he could see the impact this would have on Alberta. This obviously put Alberta in a different position on the world energy scene."

Mr. Smith now calls it "a defining moment. We looked at it and said, 'We've got something we can tell the world about.'"

Soon, the province and the Canadian Association of Petroleum Producers had launched an aggressive lobbying campaign and persuaded the influential Oil & Gas Journal to adopt the 175-billion-barrel figure in its year-end review. (Journal editor Bob Tippee now says the move made sense, if only because "at the time, we were ascribing almost zero reserves to an area then producing close to one million barrels a day.")

It didn't hurt that 9/11 and the looming war in the Middle East made the Americans keen to demonstrate that they were not entirely beholden to Mideast crude. Peter Tertzakian, chief energy economist for Calgary's ARC Financial and author of A Thousand Barrels a Second, says there is "no question" that the new reserve estimate "was a catalyst for comforting the Americans."

Spencer Abraham, now a political consultant in Washington, agrees — although he insists the decision that led to his 2003 letter to Mr. Smith "was not my call" as energy secretary. Rather, a semi-independent agency within his department crunched the numbers to ensure that politics played no part. "It was a very objectively determined conclusion," he contends, acknowledging that his vote of confidence in Alberta's resources "helped send a signal" to investors to take another look at the area.

It also signalled that the world was not running out of oil, Mr. Abraham adds. Even better: "When you have all the geopolitical uncertainties that the world of energy faces, it's great to have greater sources that are not only nearby, but also part of a country and government with whom the United States feels such closeness and affection."

In addition, the price of oil began to rise, removing much of the doubt that mining the sands could be profitable. Striving to replace dwindling conventional reserves, energy companies from China, France, Norway and Japan came hunting for a share of Alberta forest and the United States began to overhaul pipelines and refineries to handle the promised growth in Canadian exports.

By the end of last year, leases for 6.5 million hectares had been granted (the province issues more every two weeks) and Fort McMurray had become the new Dawson City, with a population that has jumped from 35,000 in 1987 to an estimated 65,000 today, not counting the army of workers who live in surrounding camps.

Fourteen companies are producing at least 5,000 barrels a day (and often far more) at 24 different sites and 30 other projects are approved or under construction. This year, the Canadian Association of Petroleum Producers predicts, companies will spend $16-billion in capital costs. They can afford it: From 2003 to 2006, annual revenue from the oil sands doubled to more than $23 billion.

Not only Big Oil is getting rich. The boom also has made fortunes for small-scale entrepreneurs quick to stake a claim.

Greg Schmidt, then CEO of Deer Creek Energy, watched as the Calgary-based company's $50-million lease at Joselyn, north of Fort McMurray, sparked a bidding war. Two years ago, French oil giant Total SA spent $31 a share (a year earlier, the price was $9.50) to snap up the company and propel its overall value to $1.7-billion. Mr. Schmidt has moved on to another project and declines to divulge just how much he took home from the deal. "I would just say," he chuckles, "the Deer Lake transaction was pretty positive to my net worth."

Of course, the projects are now so huge that some of the payoff is spreading across the nation. As Ontario Premier Dalton McGuinty said in a recent interview, "If Alberta is doing well, that's something that Canadians everywhere should celebrate. Ultimately, we all stand to benefit."

Skilled tradespeople are certainly reaping the benefits, and Dave Bohonis, a 29-year-old journeyman carpenter from Thunder Bay, is a quintessential man of the boom. He is a new father with a mortgage and willing to venture 2,600 kilometres from home to make big money.

It's also a sign of the times that his employer, Tom Jones Inc., has formally joined a dozen other Thunder Bay contractors and manufacturing firms in a bid to profit from a resource three provinces away — and the city now has its own oil-sands pitchman in Calgary.

For two weeks at a time, Mr. Bohonis and more than a dozen other Tom Jones personnel work straight 10-hour shifts to put up a new recreation centre at Suncor's Firebag site, nearly two hours drive from Fort McMurray. His nights are spent, along with about 2,000 other employees, in dorms at a work camp.

It's not a bad life, he says: Suncor pays for everything but the chips in the vending machine and serves prime rib on Thursdays. He talks to his wife, Crystal, and their one-year-old, Logan, back home every night. But he has realized – like most others in the camp, he says – that he cannot do this forever, even with a week off between his stints away. He misses his son and, although there are plans to begin direct flights, the trek from Thunder Bay this week took nine hours, with three stops, using up a day he could have spent with his family.

But oil can be habit-forming: Mr. Bohonis won't talk precise figures, but he estimates that he comes back every two weeks with nearly double what he would earn by staying home. "Everybody has a goal out there," he says. "It's always about money." And money is getting tight in Thunder Bay. Anyone who looks closely may see some irony in the fact that the closing of local paper mills is at least partly because the loonie has been driven to record heights thanks to Alberta's staggering wealth.

But one person's downturn is another's upswing. While places like Thunder Bay suffer, many Canadians enjoy the proceeds of rising oil stocks. The spotlight on Alberta ended the long-lamented discount attached to Canadian oil company shares, which have outperformed their U.S. counterparts of late. (Suncor, for instance, has become the world's best performer among big oil companies that are traded publicly.)

"So many of those stocks are held by Canadians in their RRSPs, individual stock portfolios and their pension plans," Murray Smith says, again looking back to 2003. "That recognition of the oil sands, and the telling of its story, created new wealth for Canadians from British Columbia to Goose Bay."

It's certainly a far cry from the late 1970s, when Ahmed Zaki Yamani, famous as Saudi Arabia's oil minister during the 1973 embargo by the Organization of Petroleum Exporting Countries, visited the oil sands. Consultant Robert Skinner, a former director of the Oxford Institute for Energy, says he heard later just how unfazed the sheik was while flying over the sprawling Syncrude project in a military helicopter.

"We in Saudi Arabia are very fortunate, indeed," the sheik said, stroking his trademark goatee. "We have one well that produces as much as this."

Today, even he concedes that one day the Canadian oil sands will give the members of mighty OPEC some serious competition.

Celina Harpe is 69, yet the daughter of a Chipewyan chief and his Cree bride still recalls a conversation she had with her grandfather when she was a girl. The two were standing on a bluff looking down at the Athabasca River, when the old man said: "Granddaughter, see all that water? Some day you're going to have to buy water to drink. Some day, the white man is going to come and take all our land and spoil our water."

Ms. Harpe pauses: "And today we see that. He was exactly right."

If anyone sees the oil sands as a mixed blessing, it's the bitterly divided native communities on its doorstep. Band leaders in Fort Chipewyan, a remote hamlet nearly 300 kilometres downriver from Fort McMurray, have called for a moratorium on development amid fears their cancer rate is soaring because of all the oil activity. But Fort McKay, where Ms. Harpe lives, is only two hours from Fort McMurray and its council not only endorses development, it also runs businesses that service the oil patch. Her own son drives for a band-owned outfit and she has watched young people in the community of 1,200 make enough money to buy big homes and new trucks.

A lonely voice of opposition, she worries there is a link between the rising fortunes and an epidemic of alcohol and drug abuse. She also feels that extracting the oil has poisoned a river her family once relied on for drinking water. The oil companies insist their impact on the Athabasca is relatively modest, compared with the farmers, foresters and towns that have grown up along the river. To which Ms. Harpe replies: "I tell them, 'I might look stupid, but I'm not stupid. I know what it was like 50, 60 years ago and what it's like now.' I tell that them that, from the time Suncor started more than 40 years ago, we are not able to drink the water."

Now, her community sits in the middle of a dozen proposed developments and she wants "to get away as soon as I can. I don't know where I'm going to move, but I am going to move away."

She sounds like she may have to coax herself into following through, but her concerns mirror a growing uneasiness among people across the country: Have Canadians properly weighed the costs of an oil-sands boom?

There is no question that extracting, upgrading and transporting unconventional crude leaves a crushing ecological footprint. Based on current mining leases, the oil sands may transform that Florida-sized swath of forest into a massive lunar landscape – much of it unlikely ever to return to its original state. (Existing projects have already stripped roughly 460 square kilometres.) As well, the mining operations are licensed to draw 349 million cubic metres of fresh water from the Athabasca every year, twice the amount used by Calgary, a city of one million people. Some of the water is recycled, but most of the muddy leftovers, or tailings, wind up in those toxic "ponds" that are large enough to be seen from space.

By comparison, the so-called "*in-situ*" operations needed to exploit the vast majority of sand reserves, which are located deep underground, cause less disturbance on the surface and require less water. But heating the bitumen underground and pumping it up also requires much more energy and produces far more greenhouse gas.

A swelling chorus of environmentalists – and a tide of bad publicity internationally – has led to calls to slow development until proper measures can be taken. Oil companies have managed to reduce their per-barrel environmental impact by recycling water and controlling toxins from their smokestacks. But there has been so much growth, the environmental impact has ballooned anyway.

"I don't think anybody gets how big it is and how much bigger it's going to get," says Ruth Kleinbud, an outspoken naturalist who moved to Fort McMurray in the early 1980s.

Back then, a slowdown in the oil industry had residents "fighting over moving boxes" to get out of town. Almost overnight, the sleepy community nestled in a valley surrounded by old forest stretching to the horizon was replaced by a dense, hustling, housing-short boom town, encircled by just a narrow ring of trees.

"You come down into the river valley, and it's pretty, isn't it? That's all you see, the rest is completely gone," Ms. Kleinbud says. "Things are happening so fast people aren't thinking."

A study released jointly this month by the Pembina Institute, an environmental think tank in Alberta, and the World Wildlife Fund suggested that, with few exceptions, the companies are failing to reduce their overall environmental impact significantly and have been slow to spend on new, environmentally friendly technology. And the province has been criticized for not adopting strict enough regulations for such things as when companies should restrict their use of river water during low-flow winter months.

Complicating the matter, of course, is the lingering memory of 1980, when Ottawa adopted the national energy program and wound up driving investment from Alberta. As a result, any federal effort now to restrain production to protect the environment is sure to spark a battle.

But the environmental fallout has already created more than just bad publicity. California has decided to encourage "low-carbon fuel" by imposing added levies on oil from sources it considers dirty, such as the oil sands. And last month, U.S. President George W. Bush signed a law that prevents federal departments from using synthetic oil if producing it generates more greenhouse gas than producing conventional oil does.

Many projects proposed for the oil sands tout promising new technology, but much of it has yet to be proved on a larger scale. One of the more favourable approaches – known as sequestration – involves trapping carbon gases underground. It's not required yet, although recent proposals have included space to add such facilities if they ever become mandatory.

Unless improvements are made, the environmental damage will mushroom, given that production could triple over the next seven years, says study co-author Simon Dyer, a biologist and senior policy analyst at the Pembina, which has called for a moratorium on approving further projects. "We are really at the tip of the iceberg," he warns. "If people are concerned about the environment, you don't want to be around in 2015."

In the end, logistics may be the biggest damper on developing the oil sands. The rising price of natural gas needed to heat the bitumen, the limited pipeline capacity and constraints on how much water can be taken from the Athabasca River all pose serious problems. But the biggest hurdle may be a shortage of human capital.

In the years ahead, an army of skilled workers like Thunder Bay's Dave Bohonis will be needed – as many as 30,000 of them, ranging from engineers to welders, pipe fitters and electricians. And Suncor senior vice-president Neil Camarta says the province historically has trouble if worker demand tops 10,000.

His company is already looking for skilled labour to build Fort Hills, its largest project in the sands. Phase one alone calls for 7,000 trades people and 2,500 engineers – and "we're struggling right now," he says from his Calgary office. Based on projections, he says, "there just aren't going to be enough construction workers here in Alberta available to build all these projects."

Just as many labourers were brought from China to build the railway, oil companies are now recruiting abroad. But the sudden influx of newcomers is already overwhelming Alberta's social services, causing a spike in housing costs and clogging the 435 kilometres of two-lane highway between Edmonton and Fort McMurray.

The boom has led to wage inflation and worker shortages; on the plus side, fast-food managers can earn up to $60,000 annually, and last year Wal-Mart was advertising bonuses for employees who stayed for 1,000 days. Small businesses find it hard to compete for staff; it's not uncommon, Prof. Chastko says, to drive by a darkened Burger King franchise in Calgary, closed because no one wanted to flip burgers.

The bottom line

Today, Murray Smith is 58 and no longer in politics – but he still has that framed letter from Spencer Abraham as a reminder of the campaign to create an oil power from Alberta's tub of goo. "It just showed what perseverance can do," he says. "We stuck with it. … Everybody worked hard for that result."

Now a consultant in the oil patch, he has seen black gold's fortunes rise and fall in Alberta, but feels there is no real cause for concern because the fortunes of the province and the country as well as the corporate world have nowhere to go but up. According to Mr. Smith, companies are already making huge strides toward more energy-efficient technology and environmentally responsible practices – spurred on by natural market incentives to reduce costs. "That drive is on," he says. "The next generation is here already."

Perhaps, but many others are less optimistic. They argue that none of the problems facing the oil sands – and the governments required to manage them – will be solved easily. The bigger the oil business gets, the more dependent Canada's economy becomes on a single resource. Alberta's infrastructure will be strained with the arrival of more and more workers. Until companies can produce the oil without leaving a permanent shadow on the land, the protests will only get louder.

How all these issues are weighed and resolved will shape Canada's future. "We are in an era of tradeoffs," explains Peter Tertzakian, the energy economist. "And anybody who comes forward and says, 'Here's a simple solution, do this,' is not being honest."

Senior writer Erin Anderssen and global energy reporter Shawn McCarthy are members of The Globe and Mail's Ottawa bureau. Eric Reguly is the newspaper's correspondent in Rome.

Piping detail, Syncrude upgrader, north of Fort McMurray.

Suncor is now in the process of reclaiming Pond 1,
piping some tailings to another pond,
and replacing them with gypsum to consolidate the tailings.
In 2010, the surface was solid enough to
support replanting of native vegetation.

In 2009, Suncor Energy Inc. completed a friendly takeover of rival Petro-Canada in a share deal that saw Suncor shareholders ending up with a 60 per cent stake in the new company while Petro-Canada shareholders emerged with a 40 per cent stake. The deal created Canada's largest energy firm and made it the fifth-biggest in North America.

The merged entity will retain the name Suncor while maintaining the strong brand presence and customer loyalty of Petro-Canada.

Suncor upgrader at Millennium site, Fort McMurray.

Four hundred tons of oil sands are dumped into truck with only four bucket loads at Syncrude mine.

Perhaps no subject generates such passionate debate within Canada and around the world than the Athabasca Oil Sands development. The United Nations Climate Change Conference in Copenhagen ended December 19, 2009 with a political agreement called 'The Copenhagen Accord. The Canadian Association Of Petroleum Producers summed up the position of its members as follows:

- Oil sands account for 1/1000th of global Green House Gas (GHG) emissions and were not the focus of the world leaders in Copenhagen.
- In Canada, 95 per cent of GHG emissions come from sources other than oil sands.
- Effective Canadian climate policy can only be made by looking at the oil sands in the broader context of our nation's energy needs and economic interests.
- Clearly the challenge will be to develop Canada's national climate policy and action plan in a manner that finds the right balance between environmental performance, economic growth and ensuring that we have a secure and reliable energy supply.
- The oil and gas industry remains focused on improving GHG emissions performance, largely through the application of innovative technology (for example, GHG emissions from oil sands production have declined by 39 per cent per barrel from 1990 to 2008).

For more information go to: www.capp.ca

OIL SANDS AND THE ENVIRONMENT
THE PEMBINA INSTITUTE

As you approach the Syncrude lease from the west you can see the mining operations, which remove about 40 metres of earth before reaching the oil sands bitumen that runs another 40 metres deep. In 2010 the developed oil sands mining area was as big as Waterton Lakes National Park.

Oil sands deposits are a mixture of sand, silt, clay, water and about 10 to 12 percent bitumen. Extracting oil from this sticky mass occurs in two ways. Open pit mining, which strips away the boreal forests and wetlands, is used to extract bitumen that is less than about 75 metres deep. The dominant form of extraction to date, open pit mining has already disturbed 600 square kilometres of forests since 1967, when the first oil sands mine opened. The vast majority of established reserves lie too deep for mining and therefore require in situ (or in place) extraction techniques, which involve injecting steam underground to melt the bitumen and bring it to the surface. In situ development requires a dense network of roads, well pads, water and above-ground pipelines.

While development in the oil sands began around 40 years ago, expansions have accelerated in the past decade as the global price of oil has increased. Today some 84,000 square kilometres of land underlain with oil sands deposits has been leased to oil sands companies for development, including one contiguous surface mineable zone of 4,800 square kilometres—more than four times the area of Los Angeles.

Oil sands companies are nominally required to reclaim lands after development, but the industry's track record so far has been very poor. Reclamation is defined as "stabilizing, countouring, maintaining and conditioning and reconstructing the surface of the land" after the mining operations have ceased. Reclamation is not akin to restoring what was previously on the landscape, rather it is a process that restores the "productivity" of the landscape from a perspective of how it is beneficial to human uses. Alberta has no wetland policy for forested areas, so there is no requirement to compensate for the loss of wetland habitats that are permanently destroyed by oil sands mining. Even under these lenient reclamation guidelines, less than 0.2 percent of the mined area has been certified by the Government of Alberta as reclaimed.

Photo David Dodge, The Pembina Institute

Although oil sands development is still relatively small compared to its projected growth, project-specific wildlife impacts are clear. There is troubling research about the current and growing cumulative impact of industrial development, including oil sands development, in Alberta.

In 2008, for example, Environment Canada released a landmark report showing that all caribou herds in Alberta are now considered non-self-sustaining, largely as a result of cumulative development within their ranges. Caribou herds in the East Side Athabasca River range, where in situ development is underway, have declined by over 65 percent in the past 15 years. Unfortunately, the caribou ranges in northeastern Alberta that require habitat restoration are now largely slated for in situ oil sands development.

This tailings "pond" is about five kilometers long. It is located to the north of the Syncrude oil sands operation.

Photo David Dodge, The Pembina Institute

The Suncor Upgrader on the edge of the Athabasca River in northern Alberta. Oil sands development is the fastest-growing source of greenhouse gas pollution in Canada.

Photo David Dodge, The Pembina Institute

Oil sands mining produces tailings "ponds" that now cover 170 square kilometres in northern Alberta. Tailings is a toxic liquid by-product of the oil sands mining process. Tailings contaminants include naphthenic acids, polycyclic aromatic hydrocarbons, phenolic compounds, ammonia, mercury and other trace metals, which make them acutely toxic to aquatic organisms, birds and mammals.

Mine operators are required to store tailings waste on site in large containment dykes because the water is too toxic to be returned to the Athabasca River under water quality guidelines. There are currently over 840 billion litres of toxic tailings on the landscape in the Athabasca oil sands area. The reclamation of the mines' toxic liquid tailings has never been demonstrated.

One of the major concerns associated with tailings ponds is the migration of pollutants through the groundwater system, which can in turn leak into surrounding soil and surface water. There is currently a lack of publicly available information on the rate and volume of seepage from oil sands tailings ponds, despite known incidents involving tailings seepage.

Oil sands development is much more greenhouse-gas intensive than conventional oil production because, unlike conventional oil, the overburden and the actual oil sands material must be taken out with large trucks, high pressure steam is needed to "wash" the bitumen from the sand, and the sticky bitumen must be upgraded into a more usable product. Because natural gas or diesel are used for these processes, the result is higher greenhouse gas emissions on the production side. Oil sands development is the fastest-growing source of greenhouse gas pollution in Canada.

Oil sands plants and upgraders produce roughly 40 million tonnes of greenhouse gas emissions every year. In 2006, oil sands emissions accounted for 4 percent of Canada's total greenhouse gas emissions. With an increase in production, this share could increase to 12 percent by 2020. Oil sands development is the primary reason that Canada has failed to reduce greenhouse gas emissions in line with its international obligations.

Wood bison sculpture guards the entrance to the site of Syncrude's Mildred Lake mine and upgrader. It also marks the site of the first-ever, oil sands land reclamation certificate issued to Syncrude in 2008 for its Gateway Hill.

"The pace of investment and development in the oil sands has increased quickly over the past few years. As a result, more people than ever have taken an interest in the region and they are concerned about the environmental impacts of the projects currently operating, and the ones planned for the future.

CAPP understands the concern. Oil sands development has an impact on the environment. Because it's a big industry, it has a big impact – not unlike large-scale hydroelectricity or hard-rock mining projects. The industry welcomes the scrutiny. It is essential that people understand what development means, because the issues involved are important to everyone.

The four most common environmental impacts from oil sands (sometimes called tar sands) development are greenhouse gas emissions, land use, water use and tailings ponds.

The nature of the oil sands makes them more energy-intensive to produce – energy is needed to transport the earth, to break it down into smaller pieces and heat the water used in the separation process. Energy is also used in other processes such as producing the hydrogen needed to upgrade the heavy crude. All of these steps produce greenhouse gas emissions – a contributing factor to climate change.

We're working on new technologies to lower these emissions, and capture and store carbon dioxide (CO_2).

Water is used in open-pit mining for oil separation and at *in-situ* operations to make steam. Both require a significant amount of water per barrel of oil produced. As a result, the industry needs to ensure that water use is managed responsibly.

While there is much more work to do to improve water use, there has been success in this area. Imperial Oil Resources' Cold Lake operation has reduced its per-barrel water use from 3.5 barrels in 1985 to half a barrel today, by recycling more than 95 per cent of the water it uses. And some companies are using only non-drinkable water, such as salt water (brackish), in their operations.

While only 20 per cent of the oil sands is developed by open-pit mining, those operations produce tailings ponds. These ponds contain tailings (a mixture of water, sand, clay and bitumen), are large in size and are regulated by the provincial government. Tailings ponds can be hazardous to local wildlife and are an unsightly part of the landscape.

Companies are using new techniques to reduce the size of tailings ponds and the amount of water used. For example, Shell Canada Limited's Albian Sands project uses an innovative technique that recaptures more water before tailings are released, resulting in less water withdrawn from the river and smaller tailings ponds. While just 20 per cent of oil sands are produced through mining, the tailings ponds created are large and impact the landscape.

For oil sands mines, planning to restore areas is done before the first shovel of earth is moved. Returning a tailings pond to a sustainable landscape takes many years, because the fine silts in the ponds settle to the bottom very slowly. Once settling occurs, water is removed and sent to another area of the mining operation. When the pond is properly drained, it is contoured, and topsoil, plant vegetation, trees and shrubs are replaced. Then the soil and vegetation are assessed on an ongoing basis to ensure the goals of the original plan are being achieved. When the provincial government determines the area has met their criteria for reclamation (land that is restored to a sustainable landscape), they will certify it and the land will be officially returned to the Province. Although no tailings ponds have been reclaimed to date, Suncor Energy Inc. is working to have its first pond reclaimed in 2010."

Canadian Association of Petroleum Producers

Aerial view of portion of Syncrude's 104-hectare parcel of reclaimed land known as Gateway Hill. The wood bison sculpture (photo at left) can just be seen in the bottom right corner of this image.
Photo courtesy of Syncrude Canada Ltd.

"Oil sands operations, especially open-pit mines, disturb a large area of land. In March of 2008 Alberta issued the first-ever, oil sands land reclamation certificate. Once a Syncrude Canada mine, the site was transformed into forested area. The Province designated this "a rolling forested area with hiking trails and lookout points" as the first piece of oil sands land to be reclaimed. The certificate covered the 104-hectare parcel of land known as Gateway Hill approximately 35 kilometres (22 miles) north of Fort McMurray.

Under Alberta's reclamation standards, companies must remediate and reclaim Alberta's land so it can be productive again. Alberta requires reclaimed land to be able to support a range of activities similar to its previous use.

The site was used for placement of overburden material removed during oil sands mining. By the early 1980s, the area was no longer needed and Syncrude began to replace topsoil and plant trees and shrubs." www.capp.ca

A hot water hose connects to the separator, near the bottom of this early prototype machine used to separate or extract bituminous oil from oil sands.. The combination of the blades in the machine and the hot water break up the lumpy oilsands causing the oil to float to the top and the sand settle out at the bottom of the machine. This machine was used by Karl A. Clark, an Alberta Research Council chemist in the early 1920s. It was Clark's internationally-known hot water extraction process that provided the foundation for today's $5-billion oilsands industry.

Early prototype machine used by Karl Clark of Alberta Research Council on display in AITF's lobby.

Syncrude upgrader, north of Fort McMurray.

182

Bucyrus' 495HF Electric Rope Shovel
- Gross working weight: 1,315,000 kg (2, 900,000 lbs)
- Overall height: 20.72 metres (68 feet)
- Overall width: 13.01 metres (42 feet 8 inches)
- Overall length: 28.85 metres (94 feet 8 inches)
- Single pass loading of 100 tons
- Cost: $15 million

This photo illustrates the incredible size of the machine's shovel, capable of single-pass loading of 100 tons.

Bucyrus Erie 495HF Shovel.

"The oil sands are still a tiny part of the world's carbon problem. They account for less than a tenth of one percent of global CO_2 emissions."
National Geographic, March 2009

Shovel finishing up loading truck on the right as a second one stands by for another 400 tons at Syncrude's Mildred lake mine near Fort McMurray.
Hard to believe that the reclaimed land opposite, looked just like the scene above not so long ago.

Images of oil sands development are inevitably striking. But looking at the larger picture – and understanding the new technologies that are shaping the future of the oil sands – creates a broader understanding of why this resource is important to Canadians and what the industry is doing to develop it responsibly. What is missing in National Geographic's (March 2009) photographs of the boreal forest before and during development is the "after" picture. What readers do not see is that all oil sands developments are ultimately reclaimed and returned to a natural state. The magazine provided no images of reclaimed sites, even though full reclamation is required by law and the reclamation process is already well underway. Companies must submit detailed reclamation plans to government regulators prior to starting their projects and must pay a deposit into a government administered reclamation fund over the project's life. To date, more than 65 square kilometers (25 square miles) have been reclaimed.

This is a portion of the Canadian Association of Petroleum Producers' (CAPP) response to the factual but – some industry experts believed – a one-sided and incomplete view. For CAPP's complete response go to:

http://www.capp.ca/aboutUs/mediaCentre/CAPPCommentary/Pages/NationalGeographic,March2009Issue.aspx#RlDi8mG0uieQ

For the complete National Geographic article go to:
http://ngm.nationalgeographic.com/2009/03/canadian-oil-sands/kunzig-text

Syncrude established a small bison herd on reclaimed land in 1993. Today, the herd numbers about 300 head and roams on 700 hectares of pasture land which represents about 15 percent of the total mining land that Syncrude has reclaimed to date.

LEARN TO LOVE THE OILSANDS

By Licia Corbella, Calgary Herald – December 31, 2009

Today, let's have some fun and play fairy godmother to Quebec. Let's grant the province the wish it articulated in Copenhagen. Wave the magic wand and poof, wish granted. Shut down Alberta's oilsands, except, since it's Quebec making the wish, we have to call it tarsands, even though it's not tar they use to run their Bombardier planes, trains and Skidoos.

Ah, at last! The blight on Canada's reputation shut down. All those dastardly workers from across Canada living in Fort McMurray, Calgary and Edmonton out of jobs, including those waitresses, truck drivers, nurses, teachers, doctors, pilots, engineers etc. They can all go on employment insurance like Ontario autoworkers and Quebec parts makers!

Closing down Alberta's oil industry would immediately stop the production of 1.8 million barrels of oil a day. Supply and demand being what it is, oil prices will go up and therefore the cost at the pump will go up, too, increasing the cost of everything else.

But lost jobs in Alberta and across the country along with higher gas prices are a small price to pay to save the world and not "embarrass" Quebecers on the world stage. Not to worry though, Saudi Arabia, Libya and Nigeria can come to the rescue. You know, the guys who pump money into al-Qaeda and help Osama bin Laden target those Van Doos fighting in Afghanistan. Bloody oil is so much nicer than dirty tarsands oil.

Shutting down the oilsands will reduce Canada's greenhouse gas (GHG) emissions by 38.4 Mt (megatonnes). Hooray! It's so fun to be a fairy godmother! While that sounds like a lot, Canada only produces two per cent of the world's man-made GHGs and the oilsands only produce five per cent of Canada's total emissions or 0.1 per cent of the world's emissions. By comparison, the U.S. produces 20.2 per cent of the world's GHG emissions – 27 per cent of which comes from coal-fired electricity.

The 530 sq.-km. piece of land currently disturbed by the tarsands (which is smaller than the John F. Kennedy Space Center at Cape Canaveral, Florida at 570 sq. km) must be reclaimed by law and will return to Alberta's 381,000 sq. km. of boreal forest, a huge carbon sink. Quebec, of course, has clean hydro power, but more than 13,000 sq. km. were drowned for the James Bay hydroelectric project, permanently removing that forest from acting as a carbon sink.

But fairy godmother is digressing. While the oilsands only produce five per cent of Canada's GHGs, it contributes much more to Canada's economy, with oil and gas making up one-quarter of the value on the TSX alone.

Alberta is also the largest net contributor per capita by far to Confederation. Quebec hasn't made a net contribution to the rest of Canada for a very long time. This is not to be critical; it's just a fact. In 2007 (the last year national figures are available), Alberta sent a net contribution of $19.49 billion to the rest of Canada or $5,553 per Albertan – more than three times what every Ontarian contributes at $1,757. Quebecers, on the other hand, each received $627 net or a total of $8 billion, money which was designed to help "equalize" social programs across the country. Except, that's not what's happening. Quebec has more generous social programs, like (nearly) free university tuition (paid for mostly by Albertans) and cheap provincial day care (paid for mostly by Albertans). But in this fairy godmother world, poof! those delightful unequal programs have now disappeared! Quel dommage!

The July 2009 Canadian Energy Research Institute (CERI) report states that between 2008 and 2032, the oilsands will account for 172,000 person-years of employment in Ontario during the construction phase, plus 640,000 for operations over the 25-year period. For Quebec, the oilsands will account for 84,000 person-years of employment during the construction phase, plus 292,000 for operations over the 25-year period. In total, the tarsands are expected to add $1.7 trillion to Canada's GDP over the next 25 years.

Wave wand. Poof. Jobs, gone! So, now that the oil industry has shut down and left Alberta, Alberta has become a have-not province and so has every other province. Equality at last! Hugo Chavez will be so pleased. Meeting our Copenhagen targets suddenly looks possible, as most of us can't afford to drive our cars or buy anything but necessities, so manufacturers have closed their doors and emissions are way down.

Fairy godmothers always like to look on the bright side. Quebecers finally realize they can't thrive without the rest of Canada. Alas, in Alberta, separatist sentiment has risen dramatically, citizens vote to separate and the oil and gas industry returns. Albertans start to pocket that almost $6,000 for each person that used to get sent elsewhere and now their kids get free tuition.

Fairy godmother's work is done. Quebecers must now sign up for foreign worker visas to work in Alberta to send their cheques back home so junior can start saving up to pay for college.   lcorbella@theherald.canwest.com

This article first appeared in the Calgary Herald and was used with permission of Canwest Publishing Inc.

CANADA PRODUCES TWO PER CENT OF THE WORLD'S MAN-MADE GHGS
AND THE OILSANDS ONLY PRODUCE FIVE PER CENT OF CANADA'S TOTAL EMISSIONS OR
0.1 PER CENT OF THE WORLD'S EMISSIONS.
BY COMPARISON, THE U.S. PRODUCES 20 PER CENT OF U.S. GHG EMISSIONS FROM ITS
COAL-FIRED ELECTRICITY GENERATORS AND THIS
ACCOUNTS FOR 5 PERCENT OF WORLD EMISSIONS.

Syncrude upgrader, Mildred Lake facility, Fort McMurray. Buses are taking the administrative staff back to Fort McMurray at quitting time.

Green horizons

Mark Anderson And Joanna Pachner, National Post
Published: Tuesday, May 04, 2010

There are no near-term panaceas when it comes to reducing carbon emissions in our oil and electricity-generation industries. But companies continue to move forward. They are developing technologies that may play important roles in greenhouse gas reduction in the not-so-distant future. Here, we present four: three ideas for carbon capture and storage, and a fresh take on cleaning up oil-sands tailings.

NEXEN INC. EXPLORING ECONOMIC USES FOR CAPTURED CARBON

"There are a lot of smart people looking at uses for pressurized CO_2 right now."

Calgary-based oil giant Nexen Inc. is at the forefront of the race to develop industrial-scale carbon capture and storage (CCS) systems. That's because Nexen's proprietary gasification technology allows for the capture of near-pure streams of carbon dioxide, which can then be pressurized into liquid CO_2 and stored either temporarily or sequestered permanently in underground reservoirs.

If brought into widespread industrial use, CCS has the potential to radically reduce the carbon footprint associated with oil-sands extraction and processing, by eliminating massive amounts of carbon dioxide that would otherwise enter the atmosphere as greenhouse gases.

Unfortunately, "if" is the operative word. "No one's going to do carbon capture and storage unless there's a regulatory regime in place governing CO_2 emissions, and there's a significant jump in the price of carbon," says Wishart Robson, climate-change adviser to Nexen CEO Marvin Romanow. "CCS is one of the most expensive options for meeting greenhouse-gas emission targets."

All that could change, however, if liquid CO_2 had a commercial value over and above its value in carbon credits – in other words, if there was something that could be done with the stuff other than burying it in the ground.

Funny you should ask, says Robson. "Nexen has a bunch of very, very smart people looking at different applications for pressurized CO_2 right now." Those include everything from enhanced oil-recovery technology to the production of various aggregates, chemicals and products. "That's one way of closing the gap between the cost of CCS and the current price of carbon, which, as it stands, is an order of magnitude apart."

One of the most promising uses for liquid CO_2 may be as a replacement for water as the primary solvent in the extraction of bitumen from Alberta's oil sands. Currently, it can take as many as 12 barrels of water to leech a single barrel of bitumen from the surrounding sand and clay. That's putting tremendous strain on Alberta's water resources – especially the Athabasca River. But researchers at the University of Alberta have been experimenting with pressurized CO_2 instead of water, and have discovered it does a great job of dissolving bitumen. Moreover, once the liquid CO_2 has done its job, it can be turned back into a gas and siphoned off, leaving pure bitumen.

The technology is likely still years away from commercial development, but if and when it does go into production, it could provide Nexen with another commercial market to underwrite the cost of its carbon capture and storage systems.

Project pioneer will be an indicator of whether the cost of CCS can be lowered to economically viable levels

Coal powers 45 per cent of global electricity production and, for all the work on the renewable energy front, it's expected to maintain its dominance well into the future. That means making the process of burning coal cleaner is a priority for the power industry in reaching carbon-emission reduction targets.

On that front, Calgary-based power-generator Trans-Alta is now testing a carbon capture and storage process at its Keephills 3 plant, west of Edmonton. Dubbed Project Pioneer, the effort will use technology developed by the French conglomerate Alstom, cooling the flue gas produced by coal burning with chilled ammonia, scrubbing out much of the CO_2 and then transporting it for storage in underground reservoirs. The CO_2 can then be used in the extraction of hard-to-reach oil reserves.

The project aims to develop Alstom's technology to a commercial level in hopes it can compete economically with other clean-power options, such as nuclear plants. American Electric Power is already using a prototype of the Alstom process in a West Virginia plant, but at the Keephills facility, the CCS technology – once it's fully on-line in 2015 – promises to capture and sequester a million tons of carbon, or 10 times as much. In fact, if it reaches that target the plant alone would account for 20 per cent of Alberta's target for reducing greenhouse-gas emissions.

But integrating the CCS process into such a massive facility will be challenging, says Don Wharton, vice-president of sustainable development at TransAlta. "A lot of energy is needed to run the process. It's like putting a catalytic converter on the back of your car. It has to work with all the other elements. It takes some energy." And it's not cheap. To cover the project cost, Edmonton-based utility company Capital Power will be a co-owner of the Keephills plant. The Alberta government is also investing $436 million and Ottawa is kicking in funding from its $1-billion Clean Energy Fund.

The partners are not expecting this project to be profitable. Wharton pegs the cost at $90 to $100 per ton of CO_2 captured. "We have a goal to have that cost down to $50 per ton, and that won't happen in this project," he says. "But [Project Pioneer] will be an indicator of whether it is possible."

The $50 mark is significant because that's expected to be the market price of carbon credits in 2020, when Alstom's technology is commercially available. "If [the cost] does reach $50, this will be a good investment, because it will be cheaper than buying offsets or changing fuel" to power the plants, Wharton says.

"Gasification" may be the most effective way to reduce emissions in coal-fired power plants.

There's more than one way to clean coal and extract carbon dioxide, and there's a bit of a race to discover which approach will provide the best results at the lowest cost. While TransAlta's Project Pioneer will scrub out emissions from the exhaust after the coal has been burnt, its partner in that project, Capital Power (CPC), is also exploring a technology that cleans coal before combustion.

The process, called gasification, pulverizes coal and turns it into a synthetic gas, or "syngas". David Lewin, CPC's senior vice-president in charge of the project, explains that traditional power plants burn coal to raise steam that powers turbines. With gasification, the plant is powered by the syngas along with the steam produced during the gasification process. The gas is made up of hydrogen and carbon monoxide. Adding water to the carbon monoxide turns it into carbon dioxide, which can then be compressed and piped away.

Lewin notes that capturing carbon post-combustion is inefficient because the carbon dioxide in the flue gas is highly diluted. "It's only about 12 per cent CO_2, and that makes it difficult and expensive," he says. Gasification produces highly concentrated carbon dioxide, and roughly 85 per cent of it can be extracted from the emissions. In fact, gasification produces almost no harmful emissions because the exhaust is mainly water vapour, says Lewin. The emissions of most concern – including sulfur dioxide and carbon dioxide – are eliminated or dramatically reduced. As a result, they are less harmful than those from post-combustion technology. "It's a very clean process," Lewin says.

For recently completed experimental work, CPC selected gasification technology from Siemens AG, which is already in use at petrochemical facilities in China but had never been employed in a power plant. The project also aimed to learn whether gasification could work efficiently with western Canadian coal, which is of lower grade than U.S. coal on which most tests have been done.

From that work, CPC completed an engineering blueprint for a 270-megawatt plant, but decided last fall not to proceed with construction; current electricity prices make the new plant economically unviable. (It was also hoping for support from public sources, including Alberta's $2-billion fund to promote carbon capture and storage development, but didn't meet implementation timeline criteria.) "Because of the economics, it didn't make sense to build new generation [capacity]." Lewin says. Consequently, CPC has shelved the research and its plans for now. It does, however, hope to revive coal gasification when the financial stars align – perhaps around 2020.

The industry is committed to reducing the size of tailings ponds, as well as the time needed to return them to sustainable landscapes. Companies believe research into – and development of – new technology will help them achieve this goal.

For example, Shell Canada Limited's Albian Sands project uses thickeners in the tailings that allow water to be recaptured from the tailings, before they are released into the pond. This reduces the size of the pond and the amount of water the company uses in production.

Another innovative practice is at Canadian Natural Resources Limited's Horizon project. Carbon dioxide (CO_2) is captured from the facility and mixed with silts in the tailings, which causes a reaction that forms a solid, and allows the silts to settle more quickly. This process has multiple benefits: the CO_2 is permanently trapped in the silts, and most of the water can be recycled while it's still hot, so less energy is needed to reheat it. This results in reduced greenhouse gas emissions and smaller tailings ponds.

A number of universities and agencies are also researching new methods to speed up the separation of water and silts, faster recycling of water and new ways to return the land back to a sustainable landscape.

SUNCOR ENERGY INC. MAKING TAILINGS RECLAMATION FASTER

"We won't have to build more tailings ponds and it will help us reclaim existing ponds."

Hundreds of years. That's how long it takes fine clay particles in oil sands tailings ponds to settle out and become dense enough to support overburden: the 50 metres of sand, clay, dirt, muskeg and trees that must eventually be packed back into the pond, in order for the area to be reclaimed and returned to its original condition.

Even the layer of so-called mature fine tailings (MFT) that settles out after three years is only about 30 per cent solid, the consistency of yogurt – again, too soft and spongy to support weight. "You can't put overburden on top of MFT, so you can't reclaim, and it also takes up volume so you have to build more and more tailings ponds," says Bradley Wamboldt, manager of Tailings Reduction Operations at oilsands heavyweight Suncor Energy Inc. In the 1990s, Suncor developed a method of treating tailings with sand and gypsum to speed up the release of water and firm up MFT, but getting the mixture just right is notoriously tricky, and still requires 40 years before reclamation can take place.

Now Suncor has unveiled a new technology whereby MFT is dredged from the bottom of tailings ponds, mixed with a polymer flocculent – causing clay particles to clump together – and spread on the "beach" that surrounds tailings ponds. With the clay particles adhering to one another, water is released for reuse in operations, and the MFT dries within weeks, after which it can be planted, or removed for reclamation activities elsewhere.

"It means we won't have to build more tailings ponds, and it'll help with the reclamation of our existing ponds, reducing our overall inventory," says Wamboldt. Indeed, industrial trials currently underway at three sites indicate that Suncor's new tailings reduction operations (TRO) technology can shrink reclamation times to between seven and 10 years, from the current 40 years. TRO is currently being vetted by provincial regulators, and is expected to be approved for operations-wide implementation in 2010. Suncor has applied for patents on the process, "mostly for defensive purposes, so no one else patents it," says Wamboldt, and also because Suncor's hoping to license the technology and recoup a portion of its R&D investment, tallying in the hundreds of millions of dollars.

Tailings Pond at Suncor's Millennium mine that is utilizing the new tailings reduction operations (TRO) technology described above.

Suncor's new tailings reduction operations (TRO) technology can shrink reclamation times to between seven and 10 years, from the current 40 years.

Indeed, tailings ponds already cover 130 square kilometres of Alberta's boreal forest and, in the absence of new technology, could almost triple to 310 square kilometres by 2040, according to University of Alberta estimates. "It's an issue common to all oil sands players. If our technology can help other companies deal with their tailings, it'll be good for the industry, " says Wamboldt.

Tailings Pond, Suncor upgrader, Fort McMurray.

Mark Anderson And Joanna Pachner, National Post
Published: Tuesday, May 04, 2010

Web sites used in the creation of this book include:

Archival Sources:
www.collectionscanada.gc.ca – Library and Archives Canada
www.glenbow.org – Glenbow Museum Archives
www.hermis.alberta.ca – Provincial Archives of Alberta
www.sciencetech.technomuses.ca – Canada Science & Tech. Museum
www.soulofalberta.com – Soul Of Alberta

Alberta Government and NGO Sources
www.agric.gov.ab.ca
www.albertainnovates.ca
www.environment.alberta.ca
www.fortmcmurrayhistory.com – Fort McMurray Heritage Park
www.greatcanadianparks.com
www.LeducNumber1.com – Leduc #1 Energy Discovery Centre
www.oilsandsdiscovery.com – Oil Sands Discovery Centre
www.oilweek.com – Oilweek Magazine
www.pembina.org
www.turnervalley.ca

Industry Sources
www.canwea.ca – Canadian Wind Energy Association
www.capp.ca – Canadian Association of Petroleum Producers
www.cepa.com – Canadian Energy Pipeline Association
www.growingpower.com
www.huskyenergy.com
www.imperialoil.ca
www.mackenziegasproject.com
www.nabors.com
www.northerngateway.ca
www.nov.com – National Oilwell Varco
www.oilsandsdevelopers.ca – The Oil Sands Developers Group*
www.performancewell.ca
www.petro-canada.ca
www.psac.ca – Petroleum Services Association of Canada
www.shell.ca
www.suncor.com
www.syncrude.ca
www.transcanada.com – Transcanada Pipelines

*For an excellent,
interactive map of
the oil sands click
on the map icon
on this home page: www.oilsandsdevelopers.ca

Pumpjack at sunrise, Devon, Alberta.